The Atoms Within Us

The Atoms Within Us

Revised Edition

Ernest Borek

New York
Columbia University Press
1980

Library of Congress Cataloging in Publication Data

Borek, Ernest, 1911–
    The atoms within us.

    Includes index.
    SUMMARY: An overview of human molecular biology,
with a history of discoveries and a discussion of
present developments.
    1. Biological chemistry. [1. Biological
chemistry] I. Title.
QP514.2.B67   1980   574.19'2   80-19010
ISBN 0-231-04386-4
ISBN 0-231-04387-2 (pbk.)

Columbia University Press
New York   Guildford, Surrey

To Debbie and Ronnie

Preface to the Revised Edition

Since the reading public—and the Thomas Alva Edison Foundation—have received the first edition of this book so kindly, I thought a revision which would include information gathered during the past two decades might be in order. The body of knowledge which is designated as biochemistry or molecular biology has increased explosively during that period. We have negated Wordsworth's dictum that "Science moves but slowly, slowly/ Moving on from point to point." The reason for this was the enthusiastic support of the American people of science and scientists.

Careers were offered to young people from impecunious families and thus we tapped all the talent available among us. As I shall point out in the last chapter, the mood of congressmen is changing to disenchantment because we have not cured all the diseases Pandora released upon us. Therefore, perhaps it is just as well that I summarize and present to my readers what has been accomplished in the past, as we are reaching a plateau in gaining new insights.

I have included the highlights of accomplishments in molecular biology, though the latter term is unfortunate because essentially everything within a cell studied with the tools of chemistry is biochemistry.

The reader may find that some areas in the book have not been extensively revised, but then it must be remembered that I tried to present the history of the growth of my science and the history has not changed. Above all, historians should not meddle with History.

I am grateful to Mr. Robert Tilley, at one time of Columbia University Press, who had urged me to write the original *Atoms Within Us*, and to Dr. Vicki Raeburn, editor at Columbia University Press, who also provided encouragement and support. My secretary, Ms. Ginny Wentz, provided outstanding technical assistance.

E.B.

Placer Valley, Colorado
June 1980

Preface to the First Edition

Every man, woman, and child inhabiting this planet became an unwitting guinea pig in a vast biochemical experiment on July 16, 1945, when the first atom bomb was successfully detonated. This experiment, whose outcome—the effect of an increment of radiation on life—may not be known for centuries, serves to emphasize the profound involvement of every one of us in biochemistry. Actually no such dramatic reminder should be needed. Every living creature is a chemical conglomerate existing in a vast sea of chemicals.

Biological chemistry is the study of the molecular structure and the molecular mechanism of living things. It is at present the most rapidly growing and undoubtedly the most spectacular of the biological sciences. I have written this book to acquaint the interested reader with biochemistry—my field.

I use no chemical formulas in this book. After all, they are merely shorthand symbols for facts and ideas. The formula H-O-H merely states that a water molecule is made up of two atoms of the element hydrogen and one of the element oxygen and tells something of the energy binding the atoms together. It no more depicts the exact appearance of a water molecule than a blueprint depicts the exact appearance of the engine which will be built from it.

While describing the work of the biochemist without using his baffling symbols, I tried to follow the precept for the exposition of a science to a lay audience given by the great physicist James Clerk Maxwell: "For the sake of persons of different types, scientific truth should be presented in different forms and should be regarded as equally scientific, whether it appears in the robust form and vivid coloring of a physical illustration, or in the tenuity and paleness of a symbolical expression."

I should emphasize that while I have made every effort to achieve "scientific truth," I have not attempted to present the development of the various areas of the biochemist's endeavor in complete historical sequence. I feel that an approach which must undoubtedly yield a very long and tedious book could not be risked in a work which is aimed not at the captive audience of students but rather at the free public. In most of the chapters that follow the historical background is sketched in only to serve as a frame within which a pattern of current concepts could be woven. In turn, only such current biochemical work as could be fitted into that pattern is included. What I have tried to achieve is a connected story, which shows the biochemist at his bench, traces the growth of ideas that guide his hands, and unfolds his view on the mechanism of the living machine.

I am not an admirer of the genre of science reporting in which the work of the scientist is but a thin filling sandwiched between soggy chunks of elaborate accounts of the real and imaginary drama of his life. ("A haggard man in white is pacing in the dim laboratory. His eyes, granules of shining coal in dark pits; his mouth a thin, tightly drawn ribbon as he mutters: 'This time it must work!' ") This is corn which has turned rancid with overuse. But even if I were an admirer of this sort of thing I could not perpetrate it in this book. Biological chemistry is a young science; many of the scientists whose work is included in the book are very much alive. Indeed, I have concentrated on the science rather than on the scientist to the point where I have not identified by name many of the contemporary biochemists—often only their work is mentioned. I have thus managed to evade the responsibility of choosing whom to include from among fellow scientists, many of whom are acquaintances and some personal friends.

I have kept predictions and fantasies about the future achievements of biochemistry to a minimum. Even with the most limited of imaginations it is easy to engage in such pastimes. It is an indoor sport during the coffee respite for many working scientists. But such speculations should be restricted to the experts in the field. For only they know the rules of the game, only they can clearly discern the difference between science fiction and a plausible goal within experimental reach. It is not necessary to impress the world with miracles of the future; we have achieved enough miracles already to prove our worth.

Some understanding of the work of the scientist by every member of a free society is an increasingly urgent need. The burden of the support of science has been gradually shifting. The alchemist was supported by the largesse of kings and princes; with the rise of capitalism and the concentration of wealth some oil and beer barons assumed the roles of benefactors; and, now, since the federal government underwrites more than half of the scientific research in our country, the taxpaying public has become our patron.

In 1669 the alchemist Hennig Brandt isolated what he thought was some magic stuff, phosphorus—it glowed in the dark!—from the phosphates of urine. He was supported—inadequately he thought—by his patron, a German prince. About 290 years later a far more gifted alchemist, Nobel Prize winner Dr. Arthur Kornberg, put together from more complex phosphates a magic stuff, DNA, which is the genetic material of living things. *His* patron was the United States Public Health Service, which is one of the dispensing agencies of the American public's patronage.

The role of the patron of the sciences is no less hazardous than that of a patron of the arts. Shoddy but spectacular undertakings can outshout research endeavors of depth and integrity. Only the well-informed can be discerning. I hope this book will serve as a primer in my field for my patrons.

E.B.

New York
November 1960

Contents

Science is the poetry of the intellect. . . .

LAWRENCE DURRELL

I am a little world made cunningly
Of elements, and an angelic sprite.

JOHN DONNE

1. The Stuff of Life

A fertilized egg is at once the most precious and the most baffling thing in the universe. Within the thin shell of a hen's egg is locked not just a mere mass of egg white and yolk, but a promise of a beautiful creature of flesh, blood, and bones—a promise of the continuity of life.

The mode of fulfillment of that promise is the most baffling mystery within the horizon of the human mind. The mystery of the sun's energy no longer eludes us: we have converted mass into energy; we perpetrate a Lilliputian aping of the sun with each detonating hydrogen bomb. But the mastery of the tools, of the energy, and of the blueprint which shapes an apparently inert mass into life is still but a thin hope.

A century and a quarter ago there was not even such a hope. In our striving for knowledge we had to contend not only with the secret ways of the universe, but often we had to surmount manmade barriers before we could approach our quest. One such roadblock in the path to knowledge was the principle of vitalism which dominated scientific thought until the middle of the nineteenth century. Scientists had explored with fruitful zest the nonliving world, but they halted with awe and impotence before a living thing. It was believed by the vitalists that the cell membrane shrouded mysterious vital forces and "sensitive spirits." It was an unassailable tenet that not only could we not fathom these mysteries, but that we should never be able to duplicate by any method a single product of such vital forces. An uncrossable chasm was supposed to separate the realm of the living, organic world and the realm of the non-living, inorganic world.

In 1828, a young man of twenty-eight unwittingly bridged that chasm. He made something in the chemical laboratory which, until then, had been made only in the body of a living thing. This achievement was the "atom bomb" of the nineteenth century. Its influence in shaping our lives is far greater than the influence of atomic energy will be.

Had there been science writers on the newspapers of that day they could very well have unfurled all the clichés of their present-day counterparts about the "significant breakthrough." But, oddly, not only were the man and his feat unknown to his contemporaries, he is practically unknown even today. The generals of the Napoleonic era—Blucher, Ney, Bernadotte—are known to many, but their contemporary, Friedrich Wöhler, who was a giant of intellect and influence compared to them, is known only to chemists.

Friedrich Wöhler was a student of medicine at Heidelberg in the early 1820s. His chemistry teacher, Gmelin, was one of those rare teachers who not only dispensed knowledge but "shaped souls." Under his guidance Wöhler left medicine and became a chemist. He more than justified his teacher's faith, for, in addition to the great discovery which shook the foundation of vitalism, we owe to Wöhler the discovery of two elements, aluminum and beryllium. After absorbing all that Heidelberg could offer, he went to Sweden to work with Berzelius, the greatest contemporary master of chemistry.

There Wöhler discovered, quite by accident, a method of making urea, a substance theretofore produced only by the cells of living creatures. So contrary to current ideas was this achievement that he published it only after numerous repetitions, four years later. How did he learn the secret of the sensitive spirits? How did he make urea?

Urea is a substance found as a waste product in the urine of some animals. It can be cajoled out in the form of pure white crystals, by the knowing hands of the chemist. Like all other pure compounds, urea has characteristic attributes which distinguish it from any other substance. Sugar and salt are two different pure compounds which superficially look alike. But the tongue tells them apart with unfailing ease. Their different impact on our discerning taste buds is but one of many differences between salt and sugar. The chemist has discovered scores of other differences. The elements of which they are composed, the temperature at which they melt, certain optical properties of the crystals, these are some of the distinguishing lines in the fingerprint of a compound, by means of which the chemist can recognize a particular one among the multitudes.

Urea happens to be made of four different elements. One atom of carbon, one of oxygen, two of nitrogen, and four of hydrogen make up a urea molecule. These atoms are attached to each other in a definite pattern, a pattern unique to a single substance, urea. The force which binds these atoms into the pattern of urea is the energy in the electrons of those atoms. A chemical union between atoms is a light, superficial affair. Two atoms meet, some of their outer electrons become entangled, and a temporary union is formed. The nucleus of the atom is completely unaffected by a chemical union. The energy involved in the making or breaking of such a union is minuscule compared to the vast energy locked within the nucleus of the atom—the monstrous nuclear, or atomic, energy.

Until Wöhler succeeded in making it, urea could be fashioned only in a living animal by unknown, awesome, animated spirits. The spirits whimsically flung out their product into the urine for reasons, it was thought, no human could fathom.

Wöhler had no intention of making urea in a test tube. He was studying a simple, undramatic, chemical reaction. He wanted to make a new inorganic compound, ammonium cyanate. He went through a variety of manipulations which 'were expected to yield the new substance. As the final step, he boiled away the water and some white crystals were left behind. But they were not the new inorganic salt that he had expected; they were the very same urea which animals excrete.

It so happens that in ammonium cyanate, the substance Wöhler had set out to make, there are the same elements in the same number as in urea. The difference is the pattern the atoms form. The pattern of ammonium cyanate was disrupted by the heat of the boiling water and the atoms rearranged themselves to form urea. (The changing of chemical structures by heat is not unusual; indeed it is an everyday household feat—a boiled egg is quite different from an uncooked one.)

Simple? It looks simple now, a century and a half later, when animated spirits have become scientific antiques and the manufacture of synthetic vitamins is a big industry. Thus, only in our hindsight, which sharpens with the years elapsed, does it look simple.

The importance of the finding was not lost to Wöhler and some of his contemporaries. While he wrote very modestly of his achievement in his four-page technical communication, he let himself go when writing to his mentor, Berzelius. "I must now tell you," he wrote, "that I can make urea without calling on my kidneys, and indeed, without the aid of any animal, be it man or dog."

And the master replied graciously. "Like precious gems—aluminum and artificial urea—two very different things coming so close together[1]—have been woven into your laurel wreath."

The homage of history was paid by Sir Frederick Gowland Hopkins, Nobel Prize winner, on the centenary of the discovery: "The intrinsic historic importance of Wöhler's synthesis can hardly be exaggerated. So long as the belief held ground that substances formed in the plant or animal could never be made in the laboratory, there could be no encouragement for those who instinctively hoped that chemistry might join hands with biology. The very outermost defences of vitalism seemed unassailable."

1. Wöhler discovered the new element aluminum in 1827 and announced the synthesis of urea in 1828.

We must not belittle Wöhler because his discovery came to him by chance. His finding, as some of the subsequent pivotal discoveries in biochemistry, were made because Wöhler and his successors had the perception, receptivity, and courage to recognize something new, something which did not fit into the confines of the framework hammered together by their predecessors. It is easy to add another peg into an existing edifice of knowledge—most workers in science hack out a career doing just that—but to demolish such an edifice and face the wrath and opposition of the pooh-bahs of science who rose to influence and power by building or extending the structure built on a false foundation, that takes vision and courage.

The term serendipity was coined by the writer Horace Walpole to describe the making of "discoveries by accident and sagacity, of things they were not in quest of."[2]

Wöhler's finding fits the definition to a T. He set out to make ammonium cyanate and on the way ended up by making urea, a feat which shook the foundations of vitalism.

Though the barrier between organic and inorganic chemistry was demolished by Wöhler, the terms have been retained, but with new meanings. Organic chemistry now embraces the chemistry of the compounds of carbon. This element is uniquely fecund in forming compounds. Over a million different compounds of carbon have been made by organic chemists. And new ones, by the dozen, are being added daily.

If one feels, as many of us do sometimes, that we are being drowned in those cataracts of compounds, solace can be had in the knowledge that even Wöhler felt inundated. Over a hundred years ago he wrote: "Organic Chemistry nowadays almost drives one mad. To me it appears like a primeval tropical forest full of the most remarkable things; a dreadful endless jungle into which one does not dare enter, for there seems to be no way out."

In that "jungle" of compounds are hidden veritable mines of

2. The origin of the word is from an ancient tale of three princes of Serendip (the former name of Ceylon) who were blessed with serendipity galore.

drugs, perfumes, and plastics. It is impossible to predict what uses a new compound may have. Sulfanilamide, DDT, four members of the vitamin B group, and many other drugs lay for years on the shelf of organic chemistry before their usefulness was discovered.

To inorganic chemistry is relegated the study of the compounds of the other elements. All of these together number only a paltry 50,000.

Encouraged by Wöhler's tremendous success, other chemists boldly set out in search of the products and components of the living cell. Their tools and training proved so apt for the task that we have learned more about the structure of life in the fifteen decades since Wöhler's achievement than in all of recorded history. Hundreds of compounds which had been pried out from the cell were duplicated[3] in the laboratory after Wöhler's fashion.

These successes changed medicine, changed nutrition, changed our way of life. Are you eating vitamin-enriched bread? Were you given large doses of vitamin K before an operation to stop excessive hemorrhage? Are you receiving injections of hormones? Has the life of a dear one been saved by penicillin? For all these bounties, thank Wöhler and the generations of chemists his deed encouraged.

Paralleling the chemist's search for the molecular components of the cell, biologists were charting the landmarks visible under a microscope in that Lilliputian universe. The universe is so small that, for example, it would take about five thousand human red blood cells to cover the dot over the letter i on this page. Yet as we can see in an idealized diagram of a plant cell in Figure 1.1, a cell is highly structured.

All cells, whether of plant, animal, or bacterial origin, are completely enclosed by a membrane which guards the physical integrity of the cell, maintaining it as an inviolate, semiautonomous unit.

3. The term *synthetic* has acquired an opprobrious meaning, indicating a poor substitute for the genuine. However, when the chemist synthesizes a known substance he makes an identical duplicate of what nature had made. Vitamin C extracted from an orange or from the vat of the chemical manufacturer is exactly the same material; it is impossible to tell the products apart by any means.

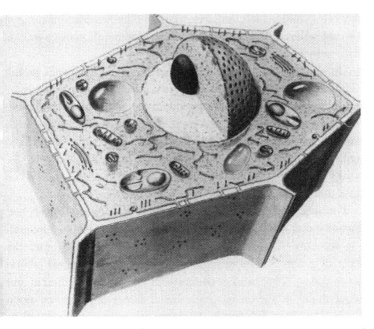

(PHOTO: M.C. LEDBETTER AND KEITH R. PORTER)

Figure 1.1. A Glimpse into a Plant Cell

Not only does the membrane guard the cell by physical containment, but it also mounts an unceasing guard over the ports of entry on the frontiers of the cell. The membrane recognizes with its uncanny molecular memory the hundreds of compounds swimming around it and permits or denies passage according to the cell's requirements. Unknown compounds which fail some subtle test for recognition on the molecular ramparts are usually excluded, and the cell is thus guarded against their possibly harmful presence.

The cell's membrane girds a kingdom of varied topography. Some of the structures, such as the organelles, which must be present in every cell of a given tissue, are obligatory. There are other structures whose distribution is more random in otherwise homogeneous cells. These are the inclusions, which may be a tiny blob of fat of a grain of starch. Organelles must be reproduced during cell division; inclusions need not be. The function of some of the organelles in the cell is well known; others still manage to hide their activities.

Our knowledge of the function of the organelles depends on our ability to remove them in more or less intact form from the cells. In order to achieve this the integrity of the cell must be destroyed. The destruction may be achieved by mechanical grinding with some abrasive powder; or it may be done by exposure of the cells to sonic vibration; or, in the case of bacteria, the cell wall can be dissolved by an enzyme, and without its protective enclosure the cell pops and disintegrates like an overblown balloon.

The next step involves a series of centrifugations at higher and higher speeds and, consequently, at gravitational forces of increasing intensity. Exposing the cells' contents to 100,000 times the force of gravity is a routine operation in most biochemical laboratories today. The various components of the cell sediment out sequentially as increasing gravitational forces are imposed upon them. The many different fractions are harvested individually; their appearance is examined under an electron microscope; visual correlations are then attempted with the organelles as they appear in the intact cell.

With luck we can obtain preparations of some organelles that retain some of their original functions. For example, there is a widely distributed structure which is just barely visible under an ordinary microscope: the mitochondrion. The reader can see a photomicrograph of a mitochondrion in Figure 1.2. We can get an approximation of their real size if we attempt to visualize a thousand such mitochondria forming a ladder across a dot on a letter i. Yet this tiny component of the cell has an elaborate structure.

The function of the mitochondrion emerged in the early 1950s as a result of two entirely different lines of investigation. On the one hand, there were investigators interested in the electron microscopic anatomy of the cell who were concentrating the mitochondria by selective centrifugations. About the same time, biochemists who were interested in the mechanisms with which the

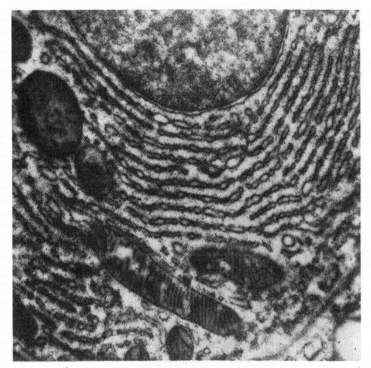

(PHOTO: JOURNAL OF BIOPHYSICAL AND BIOCHEMICAL CYTOLOGY)

Figure 1.2. The slipper-shaped structures are mitochondria.

cell generates its energy from glucose were concentrating particles from disintegrated cells that could carry on oxidation even outside the cell. When the two groups, the anatomists and biochemists, compared their preparations they found them to be the same: the mitochondrion turned out to be the furnace of the cell. It is a furnace with at least forty different working enzymes embedded in it in fixed positions. The enzymes, like so many workers on an assembly line, oxidize glucose and deposit its precious energy into a form readily usable by the cell for a multitude of tasks that require energy.

Another organelle whose function is known is the chloroplast, which is found in plant cells and in some bacteria (see Figure 1.3). The function of chloroplasts was relatively easily deduced. The tell-tale green color of the preparations indicated the presence of chlorophyll, the catalyst of photosynthesis. Thus chloroplasts are the structures on which all forms of life eventually depend. They alone have the life-generating ability to concentrate the sun's prodigious

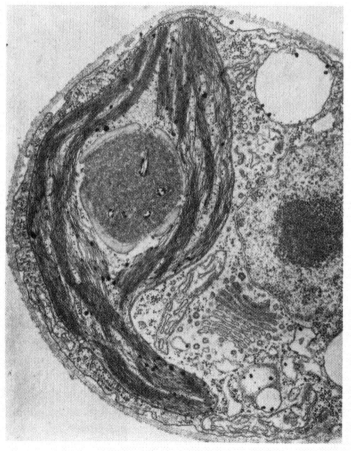

(PHOTO: G.E. PALADE)

Figure 1.3. The Chloroplast

but diffuse energy into a form which can be the fuel of life. With extraordinary efficiency they harness the electromagnetic energy of light and use it to pack random, disorganized molecules of carbon dioxide into the highly organized, energy-rich structure of the glucose molecule. In turn, the enzymes in the cells of every living organism can reverse the process and tap the energy within the glucose molecule for their own needs as they dismantle it to carbon dioxide.

Living organisms can perform various transformations of energy. The ear converts sound and the eye converts light into electrical energy. The skin can translate mechanical pressure into an electrical signal, and the nerve can transform chemical into electrical energy. Some organisms such as the firefly can use their stored chemical energy to generate light. Our muscles consume chemical energy for motion; our vocal chords do the same but produce an exquisitely controlled motion which gives rise to sound. The source of energy for all these transformations is a specialized molecular battery, adenosine triphosphate, which is produced in the mitochondrion from the chemical energy of the glucose molecule. In turn, it was the chloroplast which, with virtuoso skill, trapped the energy of light and transformed it into the energy that holds the glucose molecule together. We shall discuss this later.

Figure 1.4 is a remarkable microphotograph of an algal cell in which the product of the chloroplast, starch granules (the white popcorn shapes), can be seen emerging.

Although we have been studying the cell intensively for a century and a half, we can still discover new structures in that wondrously complex speck of life. The lysosome is a recently identified unit; it approximates the mitochondrion in size but lacks its highly organized structure. The lysosomes seem to be bags of digestive enzymes that can dismantle large molecular structures. The rubble of small fragments so produced can pass through the membranes of the mitochondrion and be consumed in its furnace.

It is obvious why such potent agents within the cell must be contained; otherwise the contents of the lysosomes would destroy the

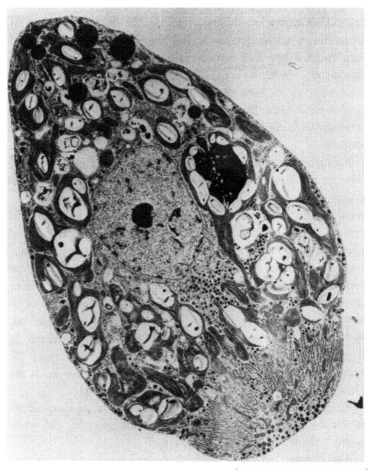

Figure 1.4. Chloroplasts at work; the white globules are starch.

cell by literally boring from within. How the structures that are doomed for dismemberment pass into the lysosomes is obscure at present. This may be the method of elimination of defunct or excess structural or functional components of the cell. The molecules which compose our bodies today will be gone in a few months and will be replaced by new ones from our foods. Even highly organized structures such as the mitochondria have but a transient life within the cell. The life span of a mitochondrion of the liver has been estimated to be from 10 to 20 days.

The reason for this profligate discarding of cellular components is now well understood. It lies in the mode of storage of biological information. Let us assume a theoretical biological molecule, say a protein, with 100 linear component units. The fidelity of its sequence of components must be absolute; otherwise it would lose its specific attributes. The information for the synthesis of this sequence of 100 components is stored in the genetic material as a single continuous entity consisting of 300 components. (The reason for this will be described later.) A protein of 100 component units can be singly cleaved in 99 different places. (The permutations of multiple cleavages is beyond my mathematical ability to compute.) Specific information for the repair of 99 cleavages with complete fidelity to the original sequence would be beyond the ability of the cell to store.

It is easier to assemble a brand new mitochondrion than to repair one whose structure has developed some molecular fissure or whose functional efficiency has faltered. An uncannily perceptive method of recognition must single out the defective structure for dismemberment by the enzymes of the lysosomes, and the resulting structural debris is fed in as fuel to the furnace of intact mitochondria.

The Golgi bodies, which were discovered in 1898 by the Italian whose name they bear, are still another example of well-defined organelles. However, despite their venerable history and ubiquitous distribution in animal cells, their function is just emerging. One hypothesis is that the Golgi bodies are the sites where some of the proteins are assembled and packaged for export out of the cell.

Very recently a heretofore unobserved organelle, the microtubule, was discovered by Dr. Keith Porter, who seems to be equally endowed with extraordinary technical skill in electron microscopy and exceptional pictorial imagination. Many electron microscopists have observed what appear to be tiny holes when a cell is sliced, removing its lid, as it were. But Porter can slice cells thinner and observe better than anyone else. In the next slice of the same cell he could observe another hole which was directly beneath the hole in the upper slice. Porter recognized that he was slicing a tube crosswise. After Porter's discovery, others developed methods of isolating intact microtubules and still others studied their method of synthesis. From these studies emerged one more sophisticated expertise of a living cell: it assembles long tubules just the way a plumber assembles them from short pieces fitted together with appropriate coupling devices. However, the analogy with plumbing ceases there. The microtubules do not seem to be conduits for fluids; their tubular shape appears to be designed to afford maximum strength with minimum weight.

What other organelles may be in the cytoplasm but too small for recognition with our present methods of observation, we do not know. Faultless structure is the hallmark of the cell, and we can, therefore, predict that in the range between single molecules and relatively giant structures such as the visible organelles there must be intermediate organizations designed for some special tasks. The detection of these "microorganelles" will be the task of future cytologists.

Recently Dr. Porter discovered such a heretofore invisible organelle. He has one of two electron microscopes in our country which can shoot electrons with a million volts of pressure through a living cell. The pictures which emerged revealed still another wondrous device nature wove to keep the components of the cell in their predetermined positions. There is a gossamer network of fine strands to which organelles and larger structures are anchored. These filaments or, as Dr. Porter calls them, trabeculae prevent the collapse of the cell's components into a chaotic mush.

Now let us come to the site of the storage of biological information, the nucleus. The nucleus is a spherical or ovoid structure encased within a membrane. Usually there is only one nucleus within a cell; however, some cells are endowed with two or more. On the other hand, not every cell possesses a well-defined nucleus. For example, whether bacterial cells contain a genuine nucleus was a controversial topic among cytologists until recently. The electron microscope resolved the argument, and it is now generally conceded that bacteria are devoid of this structure. (The bacteria do, however, have the usual genetic apparatus, although it is not packaged in a nucleus.)

During part of its lifetime the nucleus appears to be a homogeneous structure except for a small sphere within the sphere, the nucleolus, which stands out as a discrete area. The nucleus sometimes gives the impression of unchanging serenity. This period in the life of a cell was therefore called by early cytologists the "resting" stage. The description is valid only so far as visible changes are concerned. Beyond the reach of even the most powerful electron microscope there is a maelstrom of activity going on within a resting cell with explosive speed and boundless variety. The most elaborate factory with the longest assembly line is a toy compared to the cell's machinery during the "resting" stage. Every component of a living cell is being manufactured at this time at a rate sufficient to ensure that, when the time for division arrives, there will be a full stock to draw on to shape the many structures needed by the two daughter cells.

These then are the components of the living cell to which the biochemist brings the skills, tools and concepts of the chemist for exploration of that wondrous creation's anatomy at a still finer level. They are assembling a new atlas, the cell's molecular anatomy.

In this chronicle of the achievements of these chemists we may be carried away by our enthusiasm and pride in our ever-growing prowess and knowledge. We should at such times recall what the discoverer of the circulation of blood said in 1625. "All we know," said William Harvey, "is still infinitely less than all that still remains unknown." That humble statement is as true today as it was then, for three and a half centuries is a very short time to study anything as wonderfully complex as a living cell.

What have the generations of biochemists found in living things? First of all, they struck water. Lots of water. About 70 percent of the human body is water. "Water thou art and to water returnest" would be a chemically more accurate, if less euphonious, description of our corporal denouement. The amount of water in the human body is surprisingly constant. When it increases locally in a small area the tissues become swollen. There is a general increase in the water content of tissues in old age. The shrunken, externally parched appearance of old age is misleading, for the water content of the body is actually increased. Whether this increased hydration has any causal relation to aging is one of the multitude of unanswered questions which makes Harvey's humble statement all too true.

We must not find in the hydration of old age justification for replacing water as a staple beverage by more potent fluids. Alcohol actually introduces more water into the body than a similar weight of water does. An ounce of absolute, 200-proof alcohol will produce about one and one sixth ounces of water. The formation of that much water, although surprising, is perfectly possible. One of the constituent elements of alcohol is hydrogen. When alcohol is burned in the body the hydrogen is combined with oxygen that we inhale from the atmosphere, to form water.

This type of water formation is not unique to alcohol; every food is a source of water in a living creature. The camel puts this bit of biochemistry to a very practical use. To cross the desert he needs both food and water. The camel's hump, which is largely fat, provides both. Assuming that a camel's hump contains one hundred pounds of fat, the camel will derive from burning that fat huge amounts of energy—over 400,000 Calories—and, as a bonus, fifty quarts of water.

Water serves us well: it is a freight canal for the transport of foods and wastes; it regulates the body temperature by evaporating as cooling is needed; it is the remarkably efficient lubricating fluid for the body's many joints; and, finally, it makes up 70 percent of the human body. We living creatures contain only 30 percent solids.

We have about the same proportion of solids as a cup of water containing ten teaspoonfuls of sugar. Why don't we flow as freely as that sugar solution does? Why is our flesh, in Hamlet's words, "too, too solid"? Why doesn't it "melt, thaw and resolve itself into a dew"?

The biochemist's prosaic answer to the prince would be: "Because we have proteins."

Proteins are able to bind large amounts of water into themselves, forming semisolid jellies. Anyone who has ever made gelatin dessert knows this. A small amount of a dry protein—gelatin—soaks up a large volume of water and produces a semirigid mass. Not all of our proteins bind water as readily as gelatin does; we must have considerable amounts of free, unbound water in our bodies. But such water is usually confined within tubes or tissues and thus we can retain the body's characteristic solidity.

Proteins are the most characteristic components of the cell; all of life's processes are tied up in them. The term protein has been very aptly devised. It means "of primary importance."

We find proteins in every cell. Egg white, cheese, hair, and nails are composed largely of proteins. If a protein is boiled with strong acid for twenty hours it loses its identity and its characteristic properties. The edifice of the protein molecule is crumbled into its component bricks by the hot acid. The bricks that we can find in a solution of a dismantled protein are the amino acids.

There are twenty different amino acids. They all contain the element nitrogen. Hence the great need for nitrogenous fertilizers to insure good crops. Plants need the nitrogen from the soil to fashion their amino acids and proteins.

As nitrogen is the characteristic element in all proteins, phosphorous serves the same function in another large class of cellular components, the nucleic acids. (They were once thought to be restricted to the nucleus of cells—hence their name—but actually we find them throughout the cell.)

What else, besides water, proteins, and nucleic acids does the biochemist find in the cell? He finds fats and sugars, the two other large components which make up the solid matter of the cell. (All of these will be subjects of discussion in later chapters.)

In addition, the cell contains, in minute amounts, scores of other organic substances, such as vitamins and hormones. It also contains a large number of inorganic salts. Some of the salts are present in considerable amounts; of others we find but traces.

Water, proteins, nucleic acids, fats, sugars, and salts, such are the mundane substances the chemist finds within a living cell. Certainly these are not "such stuff as dreams are made on." And yet, on such stuff is built the edifice of the improbable dream that is life. All the greater, therefore, is the miracle of life.

O, it is excellent
To have a giant's strength
SHAKESPEARE,
Measure for Measure

2. Enzymes

We live because we have enzymes. Everything we do—walking, thinking, reading these lines—is done with some enzyme process. Life may be defined as a system of integrated, cooperating enzyme reactions.

The best way to get acquainted with an enzyme is to observe it in action. In order to break down a protein, say some egg white, into its constituent amino acids in the laboratory, we must use rather drastic methods. We add ten times its weight of concentrated acid and boil the mixture for twenty hours. In the stomach and small intestine the same disintegration of the egg takes place in a couple of hours at body temperature and without such strong acid. This chemical sleight of hand is performed by the enzymes made for this purpose by the cells in the stomach wall. If we add the stomach lining of a recently slaughtered hog to boiled egg white and keep the mixture at body temperature for a few hours, the egg white will be disintegrated into its amino acids just as effectively as it would have been in the stomach of the live animal.

Their enormous potency is characteristic of enzymes. One ounce of an enzyme preparation from the hog's stomach will digest 50,000 ounces of boiled egg white in two hours. The same preparation will also clot milk—it is the active ingredient of rennet powders. The potency of enzymes can be demonstrated even more impressively: one ounce will clot 2,800,000 quarts of milk.

It is astonishing how recent is our knowledge of biological mechanisms. Until the early 1820s we had no conception of what happens to the food in our mysterious interiors. Prior to that there were a host of conjectures, not the least imaginative of which was that a band of little demons was busily engaged in our stomachs, macerating our food.

Toward the end of the eighteenth century Réaumur in France and Spallanzani in Italy took the first steps toward the exploration of the stomachs of animals. They fed food, enclosed in a wire cage or in a perforated capsule, to animals and, at various intervals, retrieved the containers by means of attached strings. They noted the dissolution of the food but could not even guess at the nature of the substances which were responsible for these changes.

A whole set of new explanations was brought forth, but that they did not gain universal acceptance is obvious from the irascible fragment of a lecture by the English physician William Hunter: "Some physiologists will have it, that the stomach is a mill, others, that it is a fermenting vat, others, again that it is a stew pan; but in my view of the matter, it is neither a mill, a fermenting vat, nor a stew pan; but a stomach, gentleman, a stomach."

An accident in 1822 literally lifted the veil which covered the human stomach and its disputed processes. Fortunately, a man of rare ability was on hand to exploit this opportunity. Dr. William Beaumont was a surgeon in the recently organized U.S. Army Medical Corps. He was in charge of the post hospital at Fort Mackinac, on an island in northern Lake Michigan. The island was a busy center of fur trading. Trappers and voyageurs would swarm toward this post, their canoes laden with the winter's yield of fur pelts.

One such French Canadian voyageur, Alexis St. Martin, was part of the usual crowd at the trading post of the American Fur Company on June 6, 1822. Apparently by accident, someone's gun was discharged and St. Martin received the whole load, at short range, in his abdomen. The young army surgeon came on the double to aid the victim. On arrival, however, he realized that he could be of but little help. The bullets had ripped a huge wound, through which were protruding large chunks of lungs and what appeared to be ripped pieces of the stomach. Beaumont dressed the wound as best he could, cutting away some of the protruding flesh with a penknife, and had the patient carried to the shack which served as the hospital.

Had Beaumont had any colleagues on the post, he undoubtedly would have described the wound, between gulps of his supper, and would have expressed the doubts he had of St. Martin's chances of recovery. But Beaumont was all alone on this and on subsequent posts. His achievements are, therefore, all the more remarkable.

To Beaumont's amazement the sturdy youth survived the night. There followed a heroic struggle by Beaumont to save the patient; operation followed operation; no effort was spared to dress and drain the slowly healing wound. After several months the town officials refused to support the convalescent any longer; they were ready to ship him back where he came from—Canada—by open boat. Motivated by both charity and interest, Beaumont took the youth to his own home, where he continued to nurse and observe him.

For on St. Martin there was something to observe which had, undoubtedly, never before been seen by the human eye; an open window into a normally functioning human stomach! The gaping wound in the stomach never sealed. Its edges became healed, but a large hole in the upper part of the stomach remained open for the rest of St. Martin's unexpectedly long life. (He died at the age of eighty-three.)

Beaumont recognized his great opportunity, for he could "look directly into its [the stomach's] cavity and almost see the process of digestion." When Beaumont was transferred to Fort Niagara he took along his prize patient, who by then was ambulatory, and continued his studies.

Beaumont discovered that under the stimulation of entering food certain juices oozed into the stomach, and that the food disintegrated under the influence of these juices. He siphoned out some of the liquid and found that it could disintegrate food even outside St. Martin's stomach. There was no one with whom to share this exciting discovery. Beaumont continued his lonely studies patiently.

We must not visualize Beaumont's life as one of quiet ease or imagine that he could pursue his researches in a calm and serene atmosphere. Not only did he have his post duties, but, worse still, his guinea pig began to realize his own worth and became more and more demanding. St. Martin became tired of the diet which Beaumont fed him through his mouth and through the unnatural aperture, and he began to supplement it with liberal quantities of whiskey.

Finally, three years after the accident, he ran away to his Canadian backwoods. Beaumont was brokenhearted at the abrupt end of his engrossing experiments, but four years later St. Martin, a newly acquired wife, and two little St. Martins joined him again. In return for the support of the entire family Beaumont was permitted to resume his experiments.

He wanted to study the contents of these juices of the stomach, but, realizing that he was inadequately trained for the task, he tried to enlist the help of other physicians. He took St. Martin to New York but found, as he later wrote, that the doctors there "had too much personal, political, and commercial business on hand to turn their attention to physiological chemistry." He went to Yale, where he was advised to ship a bottle of St. Martin's stomach juices to the great chemist Berzelius, in Sweden. This he did, but there is no record of what became of it. If it did arrive in Sweden, it must have been in an advanced state of putrefaction, fit only for the slop jar.

Beaumont continued his studies on the increasingly unmanageable St. Martin, alone. This untrained, lone army surgeon, without equipment, without encouragement, and at his own expense, carried on his experiments testing the influence of hunger, thirst, and taste on the secretion of digestive juices. He antedated, by many years, the Russian physiologist Pavlov, who later repeated many of these experiments on dogs with artificial stomach openings.

Beaumont's discovery of the potency of the stomach's juices in disintegrating food, even outside the body, swept away all previous notions on the mechanism of digestion. The stomach is indeed not a grinding mill nor a fermenting vat, but an organ which can make a potent solution to dissolve and disintegrate the entering food. Several years later the first component of the juice to be identified was called pepsin. It is an enzyme which splits proteins into their constituent amino acids.

How does pepsin perform its work? As we saw in the previous chapter, a protein molecule is composed of amino acids. The couplings between amino acids are made by the shedding of water molecules, the amino acids being grafted together at the sites from which the water molecules are split. By boiling in strong acid, the forces which keep the amino acids together are broken and, at the same time, molecules of water are inserted to patch up the shorn sites. Thus the original, intact amino acids are reconstituted. Pepsin achieves precisely the same thing; it, too, breaks bonds and adds water. How it manages to do this is the most intriguing, and the most fundamental, problem facing the biochemist, for here we come to grips with the mechanism of enzyme action—and that is ultimately the mechanism of life.

How very fundamental to life enzymes are can be realized if we consider the source of energy for all living things. The sun pours vast amounts of heat and light on our earth. The sun machine rotating in the optometrist's window converts this energy, by means of its black and silver vanes, directly into motion. No living thing can do that. Living things use the sun's energy only through the mediation of enzymes. Only green plants are able, through their enzymes, to convert the sun's energy cascading upon them into a different form of energy—chemical energy. The enzymes in the cells of all other living things can, in turn, use this stored energy for their daily needs. The following chemical reaction is the pedestal on which all of life is built:

$$\text{CARBON DIOXIDE} + \text{WATER} + \text{ENERGY} \xrightarrow[\text{OF PLANTS}]{\text{ENZYMES}} \text{SUGAR} + \text{OXYGEN}$$

The 20 percent of oxygen in our atmosphere is testimony to the vast extent to which this reaction has been going on in the earth's history. All of the free oxygen in the atmosphere has accumulated from this reaction. No free oxygen could be here otherwise; it is so active in combining with other elements that, when the earth was a hot, molten "ball of fire," none of the oxygen could have escaped combination with the other elements. The reverse of the above reaction is the battery, supplying the energy for the functioning of every cell of every living thing.

$$\text{SUGAR} + \text{OXYGEN} \xrightarrow[\text{IN CELLS}]{\text{ENZYMES}} \text{CARBON DIOXIDE} + \text{WATER} + \text{ENERGY}$$

If we throw a pound of sugar into a burning stove it will burn, releasing exactly the amount of heat the sugar cane packed into it. The enzymes in our cells, however, perform this reaction in a controlled, slow process, releasing exactly the same amount of energy, gradually. (This happily slow reaction will be described in detail in chapter 4.) We are lucky that the enzymes can do it in this manner, for otherwise we would go up in a puff of smoke after a heavy meal.

Many biochemists have been attracted to the study of enzymes. Thousands of different enzymes have been discovered. There are enzymes which break down proteins and others which break down fats; some enzymes have been shown to be essential for the sending of nerve impulses; the task of still other enzymes is the building up of body tissues. In every function of the body, a host of enzymes are involved.

We believe that for every chemical process which takes place in a living cell—and there must be thousands of these—there is a separate system of enzymes. They are all remarkably specific. If there is the slightest change in the material on which the enzyme functions, the so-called substrate, the enzyme becomes impotent against it.

That enzymes can function outside of the cell was shown only about eighty years ago. This milestone is the monument to the Buchner brothers who demonstrated, in 1897, that sugar can be fermented to alcohol and carbon dioxide not only by yeast cells but also by water solutions of disintegrated cells, in the complete absence of living yeast. The name enzyme originated with this discovery. Enzyme is something *en zyme*—in yeast.

How purposeful and planned the achievement of the Buchners sounds. From reading this description of their work the reader probably visualizes the brothers working feverishly in their laboratory to prove an inspired hypothesis: that yeast juice will ferment sugar just as well as the living yeast does. It was nothing like that: they were favored by the goddess of serendipity. They were trying out extracts of yeast as a medical concoction. Since they were going to feed it to patients, they could not use the usual poisonous preservative agents. So they turned to an old wives' remedy for the preservation of their extracts. It is well known to anyone who makes jams or fruit preserves that a high concentration of sugar acts as a preservative. The Buchners added large amounts of sugar to their yeast extracts and the solutions began to ferment.

Actually, it was Beaumont who first saw enzymes working outside a living organism. Seventy years before the Buchners chanced upon their discovery he was digesting foods with St. Martin's cell-free stomach juice. The meaning of great accidental discoveries, such as that of the Buchners, cannot be recognized until the time, or rather the scientist's mind, is ripe for it. A great deal of knowledge had been accumulated in those seventy years: the paralyzing concept of vitalism had been abandoned; Pasteur explored the nature of fermentation. The whole scientific atmosphere was favorable to the search for a chemical and mechanical interpretation of living processes. Only when steeped in such an atmosphere could the Buchners recognize the meaning of their accidental discovery. As Pasteur put it: "Chance favors the prepared mind."

Since the time of the Buchners, biochemists have extracted thousands of different enzymes from a variety of different cells. Every such enzyme solution contains proteins. Slowly the suspicion grew that all enzymes *are* proteins. Before 1926 chemists were divided, however, on whether the proteins in the enzyme solutions were really the enzymes. One school of thought, particularly among German biochemists, maintained that the proteins in the enzyme preparations were impurities and that the enzymes were elusive, smaller molecules, present in minute amounts. But in that year James B. Sumner, at Cornell, was able to isolate an enzyme in a pure form, and it *was* a protein.

Sumner's achievement is so important that he was awarded the Nobel Prize for it. He had been studying the enzyme which breaks up urea into ammonia and carbon dioxide. This enzyme is called urease. (The naming of enzymes is simple and uniform: it is arrived at by adding the suffix *-ase* to the name of the substance on which the enzyme works. The scientist who first discovers the existence of an enzyme has the privilege of naming it.) Sumner chose a beautiful enzyme for his studies. The source, certain species of beans, is cheap—he grew the beans himself. The material on which the enzyme acts, urea, is also cheap. Furthermore, the enzyme produces ammonia and carbon dioxide, two of the easiest substances to assay. This in turn makes the determination of the potency of the enzyme gratifyingly simple. The more ammonia a given weight of enzyme can produce from urea, the more potent it is.

Sumner isolated urease in a pure crystalline form. How does the biochemist go about such a task? He grinds up the source of his enzyme—in this case jack beans—with water, and obtains a thin brownish soup. His obvious question is, where is the enzyme? Is it in the soluble extract, or is it in the insoluble bean grinds? To decide, he adds urea to a small portion of each. The extract promptly begins to tear urea apart into ammonia and carbon dioxide; the insoluble bean debris is impotent. The enzyme is in the extract. But how much enzyme? A painstaking measurement of the ammonia that a certain volume of the extract can produce reveals the potency of the enzyme. The chemist now has his enzyme in solution but probably hundreds of other substances must be there along with it. So he begins the long, tedious task of concentrating the enzyme. He tries to throw out of solution, by means of various chemicals, either the enzyme or some of the contaminating substances. In every case both the substances thrown out and the solution remaining behind must be tested for enzyme activity. In every case the amount of ammonia formed from urea by each new preparation must be measured. As more and more inactive material is removed, the preparation becomes more and more concentrated—and smaller amounts of it will break down larger amounts of urea. After years of work and hundreds of treatments, Sumner obtained an enzyme preparation which was a crystalline protein. It is characteristic of organic substances that they do not crystallize until they are quite pure; the impurities intrude and prevent the formation of crystals. Obtaining a crystalline, pure protein which was a highly active enzyme was a great achievement. It established that at least one enzyme—urease—is a protein.

Since then, scores of other enzymes, including pepsin, have been isolated in pure crystalline form, and every one of these, also, proved to be a protein. It took one hundred years to show that Alexis St. Martin's stomach juice owed its ability to split proteins to another protein, pepsin.

Now the biochemists know what enzymes are and what they do. But *how* do they function? How does one protein molecule—the enzyme—pry apart the constituents of another protein molecule, the substrate? Of this we know next to nothing. We are attacking the problem on several fronts, but the more we learn about enzymes the more we realize how complex is the problem and how far off the solution. Research in the field of enzymes is not unlike climbing in a strange mountain range. From a distance a peak seems near. But, as the climber proceeds, he finds hidden gullies, gigantic rock piles, and rings of smaller ridges guarding the peak, and the more he climbs, the more territory he covers, the more distant and unattainable the original peak appears.

Let us look at some of the small ridges that have been conquered in recent years. A good deal of work has been directed toward the stopping of the activity of enzymes by so-called inhibitors. We try to learn how enzymes act by learning what agents stop their action. A very small amount of cyanide stops the action of several enzymes. All of the enzymes that are inhibited by cyanide contain considerable amounts of iron. It is well known to the inorganic chemist that iron and cyanide combine into a very tightly knit compound which leaves practically no free iron in solution. Cyanide inhibits these enzymes by siphoning off their iron. (That is why cyanide is such an effective poison.) Biochemists now know that iron is essential for the action of these particular enzymes. But are we any closer to our ultimate goal? Hardly.

The study of another type of inhibition of enzymes—competitive inhibition—has not brought us much closer to the object of our quest but has yielded a whole new battery of drugs to man in his fight against bacteria: the sulfa and other drugs. (The mountaineer may never reach his peak but he may find valuable mineral deposits.)

The work of Ehrlich on the development of salvarsan is well known, but it merits continued attention because it is the source from which flowed the sulfa drugs, penicillin, streptomycin, and other aids to therapy. Ehrlich originated chemotherapy. To facilitate the recognition and classification of bacteria he subjected them to various stains and dyes. They exhibited highly individual tendencies: some were stained by one dye and not another, or, more interesting still, in some cases only part of the bacterial cell was stained.

These erratically staining bacteria guided Ehrlich in his search for new drugs. A dye stains a cell by entering into a chemical union with its contents. Since there is such a profound demarcation, even within the same cell, between staining and nonstaining areas, it is entirely possible, argued Ehrlich, that there may be some poisonous chemicals which will selectively combine with microorganisms, damage them alone, and leave the tissue cells of the host unharmed.

He had spectacular success in exorcising the parasite which causes syphilis, following his principle. He and his associates kept making arsenic-containing organic molecules, almost at random, until they hit the bull's eye with the celebrated "magic bullet," sal-varsan. This compound kills the syphilis parasite by poisoning some of its enzyme systems. Fortunately, in the doses used in therapy, it is relatively innocuous to humans.

Ehrlich's brilliant discovery nurtured the hope that new chemi-cals might be found which may be equally effective against other parasites which plague us and which are unaffected by salvarsan. The method of the search was the same as Ehrlich's—patient test-ing of each compound the researcher could lay his hands on. It was prospecting among organic compounds for new drugs, instead of in sand for gold.

There were no guiding principles in the search. Each compound was tried on test-tube culture growths of various bacteria, and if any showed promise by killing the bacteria they were tested on mice or other experimental animals. For many a drug is effective in the test tube but useless in the whole animal, either because it is too poi-sonous or because it is rendered harmless to the bacteria by condi-tions in the animal. The work is slow and tedious. For example: a hundred mice might be injected with an identical dose of a virulent strain of streptococci—the little beasts which cause "strep" throats. In addition to this injection, fifty of the mice might receive the same dose of the drug which killed those streptococci in the test tube. In transferring from the test tube to the mouse the dosage of the drug is calculated by proportion. If one milligram of the drug killed the organisms in 1 cubic centimeter (about 1 gram) of broth, a mouse weighing 30 grams would get 30 milligrams of the drug. Then all that is left to do, after an impatient night, is to count the dead mice in each group. If the majority of the drug-protected mice survive while the others perish, the drug is effective.

In 1935, twenty years after Ehrlich's death, Domagk, a German physician, stuck gold in a dye called Prontosil. The new drug passed all the preliminary tests with flying colors: it protected mice against the streptococci, and it was harmless to the mice. The first human saved by the drug was Domagk's own young daughter, who had come down with a severe case of streptococcal blood poisoning. At that time, the physician could only lance the wound where the organisms made the breach, and hope that the body's natural defenses would rally and exterminate the invading cocci already in the blood stream. When his daughter continued to sink, Domagk fed her large doses of Prontosil. She rallied and recovered.

This dramatic success was the first of a series of spectacular demonstrations of the value of the new drug. Prontosil was patented by the I.G. Farben drug cartel, and the stage was set for the world monopoly of this potent drug.

"Fortunately for the world, however, Tréfouel and his colleagues in Paris soon showed that Prontosil acted by being broken up in the body with the liberation of sulfanilamide, and this simple drug, on which there were no patents, would do all that Prontosil could do." The quotation is from Sir Alexander Fleming, the discoverer of penicillin, the production of which the English scientists made available to all—in Sir Alexander's words—"without thought of patents or other restrictive measures."

Tréfouel was a fine organic chemist whose chief at the Pasteur Institute of Paris asked him to analyze Prontosil for its components. A few days later the chief found a large bottle of sulfanilamide on his desk with a note from Tréfouel! "With my compliments."

Tréfouel's discovery—one of the few effective blows by a Frenchman against the Germans in that decade—gave tremendous impetus to search for new antibacterial agents.

Sulfanilamide is a relatively simple compound; a competent sophomore in chemistry can make it. Furthermore, the organic chemist can make variants of it with ease. Onto the structure of sulfanilamide he hangs a variety of groups of atoms, and with high hopes he hands the new products over to the bacteriologist for testing.

Hundreds of altered sulfa drugs were made; some were better than the original sulfanilamide or Prontosil, others proved useless. But the search was still hit or miss. A guiding principle, an insight into the mechanism of the killing of the bacteria, was lacking.

The English scientists Paul Fildes and D.D. Woods proposed an attractive theory. It is based on a theory of Ehrlich—the lock and key theory—which that remarkable genius had proposed for the explanation of the mode of action of enzymes. The history of the lock and key theory repeats the weary pattern of the reception of new ideas. Ehrlich was ridiculed by his contemporaries, but a new generation of scientists returned with admiration to the much abused theory and used it eagerly.

Ehrlich pondered the specificity of enzymes. Why does an enzyme act on one substance and not on another? He theorized that an enzyme and the substances it can alter must fit into each other as a key fits a lock, and that the possibility of such a union determines whether the enzyme can function on a substrate.

Fildes and Woods extended the theory to its next logical step. What would happen if another substance, which simulates in appearance the normal substrate, would fit into the lock of the enzyme molecule? The key may fit, but not completely; the enzyme mechanism may jam. They pointed to the possibility that sulfanilamide may simulate in structure some substance in the diet of bacteria. Thus, sulfanilamide may crowd out the dietary essential from the lock of the bacterial enzyme.

But what is this dietary essential? It was already known that an extract of beef liver can protect bacteria against the sulfa drugs. If, to a suspension of bacteria, liver extract is added along with sulfanilamide, the drug is made impotent; the bacteria flourish.

A hunt was started to track down the substance in the liver which fortifies the bacteria against the sulfanilamide. The suspension of finely hashed liver was subjected to a variety of chemical manipulations to determine, for example: Is the factor soluble in alcohol? Can it be thrown out of solution by adding chemicals which invariably precipitate proteins? In every case the material before treatment and each fraction obtained from the chemical manipulation had to be tested for its ability to overcome the toxic effects of sulfanilamide. Casting this chemical dragnet is dull and tedious work, but the hope of tracking down the active material spurs lagging spirits, and occasionally the researcher's patience is rewarded. Since unsuccessful searches are seldom presented to the lay reader, solutions of problems of this kind must sound monotonously simple. They are far from it. Sometimes, years of exhaustive—and also exhausting—intellectual and physical labor yield nothing, and the search is sadly abandoned. However, this search was fruitful. The substance was isolated in pure form and to everyone's amazement it turned out to be a well-known chemical compound, para-aminobenzoic acid. This laboratory reagent is the essential substance and, indeed, a vitamin for bacteria.

The chemical structures of para-aminobenzoic acid, nicknamed PABA, and of sulfanilamide are strikingly similar. The conjecture which launched the search was beautifully confirmed. The enzymes of the bacteria may mistake sulfanilamide for PABA. The enzymes receive sulfanilamide, but since it is not completely the same as PABA there is soon confusion in the cell. Enzyme mechanisms stall, the bacteria cannot grow and cannot reproduce. If an extra dose of PABA is given to the bacteria at the same time as the sulfanilamide, the bacteria are no longer overpowered by the sulfanilamide, and they continue to live. If the dose of sulfanilamide is again increased sufficiently, the bacteria once more will not grow. There is a definite numerical relationship between the amounts of PABA and sulfanilamide which determines whether certain bacteria can live or not. Growth of a certain species of bacteria may be stopped if, in the fluid where they live, there are 1,000 molecules of sulfanilamide to one molecule of PABA. Their ability to grow is restored if the PABA concentration is increased to a ratio of 1,000:2. The various sulfa drugs differ in the PABA-sulfa ratio which will stop bacterial growth. For those bacteria which required 1,000 molecules of sulfanilamide to prevent growth, 10 molecules of sulfathiazole will suffice. Therefore, a patient invaded by these bacteria needs to be dosed with much smaller amounts of sulfathiazole than of sulfanilamide.

Biochemists visualize the sulfa drugs as competing with PABA for the favors of some enzyme in the bacterial cell. Such inhibitors of enzymes and of bacteria are called competitive inhibitors. Studies of competitive inhibitors yield just a tiny glimpse of the enzymes' work. We conclude that the enzymes must combine with a substance like PABA. The combination is probably chemical in nature. There must be a preexisting "lock pattern" in the enzyme into which PABA must fit, and into which sulfanilamide also fits. However, in a later step in the process, sulfanilamide must jam the machinery.

The concept of competitive inhibitions also provides us with a rational approach to the search for new drugs. We need no longer try compounds at random. There is now a guiding principle in fashioning new "magic bullets." We try to make compounds similar in chemical structure to substances which are essential for the parasites, hoping that the new compound may be a competitive inhibitor in the parasite, without injuring the host.

Our task is not simple, however, for it is an extraordinarily fortunate set of chance circumstances that renders sulfanilamide toxic to bacteria and relatively innocuous to mammals. PABA is usually absent from the bloodstream of mammals. It is but a building block for a more complex vitamin, folic acid, which mammals must acquire from their diet. The folic acid in the blood stream of mammals is, in turn, not available to most bacteria, because it cannot cross the barrier of the bacterial cell wall. Many bacteria are thus forced to fabricate their own folic acid from PABA, and that is their Achilles' heel, where they are vulnerable to the onslaught by the drug from Paris, sulfanilamide. At the same time, the mammalian host is immune to the toxic effects of sulfanilamide, since it does not depend on its own enzyme system for the fabrication of folic acid.

Competitive inhibition is the guiding principle in the search for drugs for the therapy of cancer. Altered molecular structures have been designed to simulate every type of component ever found in the cell. Every skill of the organic chemist has been pressed into service in this vast campaign of dissimulation. Of the thousands of compounds prepared and tested several proved to be of considerable value in therapy. Unfortunately, however, the cancer cell is more than a match for the ingenuity of the biochemist. As fast as we develop new therapeutic agents the cancer cells develop mechanisms of resistance rendering the drugs impotent. Of course the ideal drug—yet to be discovered—would be one which could be selectively inhibitory only to the cancer cell as sulfanilamide is toxic only to bacteria. But in order to fashion such a drug we must know more about the enzyme systems in normal and in cancer cells.

A cell achieves its individuality and its very size by the kind and the amount of enzymes it has. Pathologists can recognize the characteristic component cells of the various tissues by the presence or absence of certain topographical features and by the interaction of cells with certain dyes. The biochemist is able to differentiate the cells of different tissues by determining the kinds of enzymes and their relative levels present. We have just started to explore the molecular mechanism which determines the levels of enzymes in a cell. As we shall see in a later chapter the potential capacity to produce an enzyme resides in the gene. But, for the maintenance of the levels of enzymes, the cell has devised ingenious regulatory mechanisms. It had to; no enzymes could be permitted to accumulate at random, otherwise the harmonious integration of cellular function would be disrupted by an anarchy of competing enzymes, leading to the cell's destruction.

One of the regulatory processes is a feedback or cybernetic mechanism. The latter term was coined by the mathematician Norbert Weiner to describe a system which is regulated by its own final product. A mechanical example of a cybernetic regulator is the governor on a steam engine. This consists of two steel spheres whose rotation is actuated by the pressure of steam in the engine. The spheres are so suspended that they rise with the centrifugal force conferred on them by increasing speed. At a predetermined speed and steam pressure a sleeve attached to the spindle of the spheres rises beyond a critical height and actuates a valve which releases the steam pressure. With the reduction of pressure the speed of the engine is slackened, consequently the governor's spheres fall and the steam valve is shut. Thus, excessive speed feeds back that information to the machine and self-regulation is achieved.

That such a feedback mechanism exists in living cells was discovered simultaneously by two different investigators, D.D. Woods of Oxford University and J. Monod of the Pasteur Institute. It has been known for a long time that if bacteria which had been growing on a complex diet containing a variety of nutrients are transferred to a scanty environment containing only sugar and salts, the bacteria will resume growth only after a long lag. They were "adapting" to their new environment. What could be the mechanism of this adaptation? It was found that the bacteria were accumulating enzymes for the fabrication of some of the nutrients of which they were deprived in their meager environment. If one of these lacking nutrients, say an amino acid, is returned to the bacterial broth the enzyme for its fabrication disappears from the newly growing cells. In other words, a system of feedback communication exists within the bacterial cell which orders the halting of the making of an enzyme since there is an abundance of the end product of the enzyme's activity.

Our understanding of the molecular mechanism of this communication system has emerged recently. We understand the mechanism only at its simplest level, the regulation of the appearance and disappearance of a single enzyme. But how feedback control operates to regulate the shape and size of organs is beyond our current intellectual reach. For instance, we can remove half of the liver of an experimental animal or of a human and it will grow back exactly to its original size. How is this achieved? Indeed, how does the whole organism arrive at its predetermined size: why does a cat not grow up into a lion? It can be confidently assumed that just as feedback systems operate to control an enzyme component of a cell, the same kind of mechanisms probably control the number of cells as well.

As some enzymes are suppressed by their end products, others can be cajoled to appear by the beckoning of their substrates. Some microorganisms cannot normally use milk sugar as a source of food. Milk sugar is a complex sugar made of two smaller sugar molecules, glucose and galactose. Some bacteria thrive on either of these simple sugars, but they will not grow if their diet contains only milk sugar. They lack the enzyme needed for the cleavage of milk sugar into its components. However, after prolonged incubation the lifesaving enzyme begins to accumulate within the starving cells; the organisms begin to grow. Such "adapted" cells can thereafter always use milk sugar, provided the dietary regime is not interrupted. If there is an interruption by the feeding of the simpler sugars, the milk-sugar-cleaving enzyme disappears. One might easily conclude that here we have an example of a lifesaving teleological mechanism. However, other substances, too, can provoke enzyme formation, provided a key pattern of a few atoms present in milk sugar is also present in the substance which duplicates the action of milk sugar and the bacteria cannot use some of these enzyme-inducing substances as food. Thus the production of the milk-sugar-splitting enzyme cannot be teleologically oriented for the feeding of the bacteria; rather, it appears to be a response to some stimulus by a certain pattern of atoms. Just how such a pattern of atoms can evoke the synthesis of an enzyme is as obscure at present as the mechanism by which the enzyme itself functions.

While our knowledge of the intimate mechanism of enzyme action is very limited, we do have much information about the different types of enzymes with which living things are equipped. Of all the enzymes, there are none more fascinating than those that produce light. Yes, "The fireflies o'er the meadow," which "In pulses come and go," owe to an enzyme the distinction of being lifted by the poet from the obscurity of thousands of other insects. They possess an enzyme which produces light. The firefly is not the only creature gifted with an astonishing flashlight. A variety of species, from the microorganisms that light up the wake of ships at sea to large deep-sea fish which have lanterns to illuminate the otherwise eternal darkness of their hunting grounds, all use enzymes for the generation of light.

The enzyme has been named luciferase by E. Newton Harvey, the biologist to whom we owe much of our knowledge in this field. The light produced by luciferase is a cold light. The enzyme is able to generate light without wasting a good deal of energy as heat. Mechanically, we are no match for the firefly. Even fluorescent lights, which are a great improvement over the hot-filament bulbs, generate some heat.

The most wonderful of these lantern-carrying creatures are certain fish which, though unable to make light themselves, carry around millions of light-producing microorganisms in little pockets under their eyes. When they want their lights dimmed, or put out, they cover the pockets with a convenient lid.

The purpose of these devices is not only illumination; the firefly has its light on its abdomen. Some deep-sea fish have light pockets along their sides, in characteristic porthole-like patterns, which vary from species to species. It is thought that members of the opposite sex of the same species recognize each other by these patterns.

Some animals use enzyme systems for attack or self-defense. The electric eel can discharge several hundred volts, over and over, to incapacitate its prey or enemy. The largest of these creatures, some of which have been captured in the Amazon river, grow to five or six feet in length. About three fourths of the body is the electric organ which generates electricity with its remarkable enzyme system.

The electric organ of the eel is crammed with the same enzyme which is found in the nerve tissues of other animals, including man. Since the nerve impulses are relayed through electrical charges, the appearance of the same enzyme in these two entirely different tissues is not too surprising. It is one more illustration of what Emerson described as the poverty of nature. "And yet so poor is nature with all her craft, that from the beginning to the end of the universe, she has but one stuff . . . to serve up all her dream-like variety. Compound it how she will—star, sand, fire, water, tree, man—it is still one stuff and betrays the same properties."

(Exception must be taken to considering nature "poor with all her craft." Indeed, she is astoundingly rich in versatility, achieving, as she does, infinite variety with "one stuff.")

The venom of snakes is another example of assault by enzymes. There are two different types of snake venom. While both of them attack the blood of the prey, their mode of action is different; one causes the disintegration of red blood cells, the other clumps the red cells together. Since intact corpuscles are essential for the transportation of oxygen to outlying tissues, the victim of snakebite will virtually suffocate for lack of oxygen.

Further explorations in enzymes will yield rich rewards. Scratch the surface of almost any problem in medicine and you expose a problem in enzymes: diabetes is a derangement of the enzymes which tackle the release of energy from sugars; cancer—a disease of unnaturally rapid and shapeless growth—is probably the result of some monstrous blunder in the functioning of enzymes; aging may yet be shown to be due to the slowing down or inhibition of some pivotal enzymes.

Whether a cut in the skin will become infected is decided, not only by the abundance of bacteria around the wound, but also by the abundance of antienzymes present in the blood. It was discovered that some bacteria produce a very useful enzyme which can dissolve the protective surface coatings of their prey. Using these enzymes as a battering ram, the bacteria can penetrate their prey all the more easily. The enzyme had originally been named hyaluronidase but this was mercifully changed to invasin. Invasin is inhibited or rendered ineffective by an antienzyme, anti-invasin, which is normally present in the blood stream but which is deficient in patients suffering from severe bacterial infections. Too little information is available to tell how these antienzyme shock troops of the body repel the invading enzyme; whether they are enzyme inhibitors or enzymes which gobble up other enzymes. Time and more research will tell.

At the time of the original discovery of invasin, the enzyme appeared to be of only academic interest. But research sometimes takes unexpected turns. Goethe's dictum "what is true is fruitful" has been often borne out in the growth of science. An apparently modest little discovery may swell through the work of the original discoverer, or of others, to impressive proportions, providing excitement and joy to all researchers—particularly to the one who made the initial observation. What may develop from studies of invasin and anti-invasin is impossible to predict now. A totally unforeseen practical benefit has already accrued from it: the remedy of a certain type of sterility in humans.

To make the appointed task of the sperm easier, the human sperm fluid is richly stocked with the very same enzyme the bacteria produce, invasin. In some cases of sterility it was found that the male is deficient in the production of invasin, and the sperm, unaided by this mighty trumpet, is impotent before its Jericho. The shortcoming has been effectively overcome in a number of cases by the appropriate use of preparations of the enzyme from bull testes.

So far, discussion has been restricted to what has already been accomplished in enzyme chemistry. But let us now take a peek into the enzyme chemistry of the "brave new world" of the future. The writer has a pet enzyme inhibitor, alas, yet to be discovered, which may solve a social problem that has distressed men of good will and of ill will for generations.

The extent of the pigmentation of the body is determined by enzymes. As one of our amino acids, tyrosine, is combined with oxygen, it forms dark-colored products, the so-called melanins, which produce the color of the hair, eyes, and skin. We see the result of the complete absence of this enzyme, through an error of heredity, in albinos with snow-white hair, fair skin that cannot tan, and unpigmented pink eyes. On the other hand, we have the more abundant pigmentation of the colored races. According to our best authorities, "No pigments other than those found in the whites are encountered in the dark races," and therefore, "the colored races owe their characteristic color only to variations in the amount of melanin present"—and, in turn, to an overactivity of the enzymes which produce melanin. The difference between the blondest and the darkest of humans is only enzyme deep.

It is entirely within the realm of biochemical possibility that someday a specific inhibitor will be found which, when fed, will slow down the enzymes which produce the melanin pigments, enabling us to lighten skin pigmentation at will. Or, on the other hand, we might find an enzyme *accelerator*, as well, which will enable us to darken lighter skins. For, after all, who is to decide what is preferable? The judicious use of an enzyme inhibitor and accelerator may thus someday achieve a Utopia of color.

. . . the little things are infinitely the most important
SHERLOCK HOLMES

3. Vitamins

Why do we need vitamins? Why would the absence of a minute dust of white powder from a sailor's diet cripple him with scurvy? What can the vitamins do in our cells to make us their slaves?

The question of the role of vitamins in the cell will be answered against the background of an acrimonious scientific controversy which raged for almost eighty years before it was completely resolved.

Solution to a problem in science does not pop out of the head of one genius, like Athena out of Zeus's forehead, fully developed and completely integrated into the rest of the body of scientific knowledge. Our understanding of the role of just one vitamin required the intellect and labor of generations of scientists.

Before we get on to the controversy let us introduce the two protagonists who started the feud. In one corner is the champion, Louis Pasteur, the greatest biological scientist of the nineteenth century, perhaps of all centuries. At the age of thirty-eight, when this controversy started, he was already a famous chemist with a list of brilliant achievements to his name. About this time he was leaving the field of the chemistry of crystals and was laying the foundations of bacteriology as a science. Later he studied the "diseases of wine" under municipal sponsorship and the diseases of silkworms under commission of the French ministry of agriculture.

From his studies, Pasteur concluded that the "diseases of wine" were produced by bacteria which contaminate the juice of the grape, and whose nefarious by-products were assaulting the French palate. He prescribed appropriate remedies: warming the unfermented juice to destroy the trespassing, undesirable organisms. Had he gone no further, Pasteur would have been universally hailed by his countrymen as a savior of his country second only to Joan of Arc. (No doubt there must have been many who would have rated delivery from the despoilers of wine even higher than delivery from the British.) But Pasteur did go further. He widened his researches and his conclusions and announced that some human diseases, too, are produced by bacteria. With this statement he came into a head-on collision with some members of the French Academy of Medicine. There ensued a series of celebrated polemics—including at least one invitation to duel—in which Pasteur showed himself not only a man of genius but also a man of iron will, ready to fight for truth as he saw it through his microscope. We get a glint of the steel in the man, in this earlier, less publicized controversy.

The fermentation of sugar into alcohol and carbon dioxide had been known for a long time, but the motivating force which induces fermentation was unknown. The German chemist Liebig had proposed the most popular theory of the time. Fermentation was supposed to be produced by the last vitalistic "vibrations" of dead biological material. (Actually, this bizarre hypothesis was proposed some hundred and fifty years earlier by G.E. Stahl, a physician to the king of Prussia.) Decades had passed since Wöhler's synthesis of urea, but vitalism still dominated the minds of many scientists.

Pasteur concluded from his own studies that, on the contrary, fermentation is the normal function of living yeast cells, and that it proceeds apace with the growth of yeast cells. He published his views in 1857 in a historic paper: "Mémoire sur la fermentation appelée lactique."

Pasteur was not the first one to make the correlation between living yeasts and fermentation. A compatriot of his, Cagniard de Latour, had arrived at the same conclusion twenty-two years earlier. Latour stated that yeasts were living organisms, "capable of reproducing themselves by budding, and probably acting on sugar only as a result of their growth." But even scientific truths must have apostles to struggle for their acceptance, and the struggle sometimes gets rough. Pasteur was exceptionally equipped for both tasks: to see truth and to fight for it. He proceeded to prove that fermentation did not require biological material "vibrating" in its moribund throes. In 1860 he published a paper showing that fermentation did not require any foreign protein. He claimed that from a solution of mineral salts, ammonium salt, sugar, and a very small seeding of yeasts, "the size of a pinhead," he obtained both fermentation—he zestfully described the copious evolution of carbon dioxide—and a lush growth of healthy yeast cells.

This was electrifying news. Here was evidence that a living thing, a yeast cell, can grow and reproduce in a medium completely devoid of any mysterious vitalistic substance. Only sugar, minerals, and ammonia were needed. Such a fundamental experiment was bound to be repeated.

And now, the challenger: Justus Freiherr von Liebig, the dean of German chemists. In 1869, at the age of sixty-six, he could look back on a life rich in achievement in organic and agricultural chemistry. Liebig announced that he could not repeat Pasteur's experiment! Furthermore, he literally insulted Pasteur by suggesting that he deceived himself by mistaking for yeast some stray molds growing in his flask. Pasteur, the greatest living expert in bacteriology and a handy man with a microscope, unable to distinguish a yeast cell from a mold filament! He replied with characteristic pungence: "I will prepare, in a mineral medium, as much yeast as Mr. Liebig can reasonably ask, provided that he pays the cost of the experiment." Furthermore, Pasteur invited Liebig to come to his laboratory so that he might repeat the experiment in Liebig's presence. Liebig was, for those days, an old man, and he died four years later, in 1873, at the age of seventy, without accepting Pasteur's defiant challenge.

The first round went to Pasteur on points.

The next important development in the controversy came in 1901, six years after Pasteur's death. Wildiers, at the University of

Louvain, calmly restudied the problem of raising yeast cells in a mineral medium. He found that the crux of the problem was the size of the droplet of yeast cells used to inoculate the sterile mineral broth. Pasteur said he used a droplet the size of a pinhead. Unfortunately that was not a very exact prescription. Just as medieval philosophers are said to have debated on how many angels could stand on the point of a pin, bacteriologists began to debate the *size* of a pinhead.

Wildiers found that if the size of the inoculating droplet was very small, yeast cells did not grow in Pasteur's medium. The few that grew were sickly looking, malformed little creatures. He reluctantly took Liebig's side in the controversy. But he went further than that. He showed that with a large droplet he could repeat Pasteur's successful experiments. Furthermore, he could use a very small inoculum of live yeast cells plus a large droplet of sterilized yeast cells, in which all the organisms were killed, and, adding these two to the mineral broth, he obtained lush yeast growth. He concluded that there is something other than live cells in the large inoculum which the yeast must have for growth. He called this something bios—from the Greek word for life. Wildiers found that bios is present in a variety of substances. A sterile extract made from meat or from egg yolk, added to Pasteur's broth, enabled the yeast to grow from minute inocula.

There was something present in these extracts which was indispensable to the growth of yeast cells. Thus Wildiers demonstrated the existence of vitamins, long before the term was coined. He tried to isolate the bios. But the current techniques of chemistry were not up to the task.

One might expect that progress in the bios problem should have been rapid after this. Far from it. The controversy really began to rage in earnest. At first it was denied that there is such a thing as bios. Experiments were brought forth showing that yeasts do not need bios. These may very well have been sound experiments, for we know today that different strains of yeasts do have different nutritional requirements. Some antibios crusaders held that bios was not essential; it merely overcame the poisoning of yeast by copper. Verifying Pasteur's famous dictum—"Nothing is so subtle as the argumentation of a dying theory"—some went even so far as to say that yeast grew better with extracts of meat, not because of bios, but because of some other substance. These bacteriologists, it has been said, were "qualified to join the Last Ditch Bacon Club, which holds that Shakespeare's plays were written, not by Shakespeare, but by someone else bearing the same name."

The bios contenders continued to wrangle; anyone who could think of nothing more productive to do could always show that some exotic substance did or did not contain bios.

This round belonged to Liebig, the challenger.

Before we go into the final round in the controversy we must turn our attention to another area in science where big strides, which eventually led to the resolution of the bios problem, were being quietly made.

The knowledge that foods can remedy some human diseases is as old as recorded history. In some Egyptian papyri we find descriptions of the ritual by means of which the priests restored the eyesight of travelers, returned from prolonged trips in the desert. With appropriate incantations they fed to the afflicted the liver of a donkey sacrificed under suitable omens in the sky. In the Apocrypha there is the story of Tobit (Tobias, in the Vulgate), who lost his eyesight. His son Tobias was instructed to catch a monster from the sea and "anoint" his father with its liver. It is difficult to determine, at this distance, how the son might have interpreted the original term for "anoint." If he fed the "monster" liver to his father he practiced perfectly sound vitamin therapy. The lack of vitamin A in the human diet causes, at first, night blindness and, later, almost complete blindness. The richest source of this vitamin is the liver of animals, especially fish.

The cod-liver oil industry ought, perhaps, to make young Tobias its patron saint, for he was the first to practice the trade. The use of fish-liver oils in therapy in modern times was first mentioned in 1782, when the English physician Robert Darbey wrote, "an accidental circumstance discovered to us a remedy, which has been used with great success . . . but is very little known, in any country, except Lancashire. It is the cod, or ling liver oil."

How very recent is our knowledge of vitamins can be appreciated from the following quotation from a leading textbook on diet. "The chief principles in food are: Proteids [archaic name for proteins], Carbohydrates [sugars], Fats, Salts, Water." Not an inkling of anything else. That book was published in 1905.

Since the story of the discovery of vitamins has been often told it needs to be summarized only briefly. A Japanese admiral, Takaki, had a hunch that the beriberi with which sailors on long voyages were plagued might be due to their poor diet at sea. He was a born experimentalist, for, in 1882, he sent out a ship well stocked with meat, barley, and fruits, and indeed, no beriberi occurred among the crew. This was clear-cut evidence for the relation between diet and disease, but of course there was still no glimpse of what was lacking in the diet.

Fifteen years after Takaki's cruise a great stride was made by a physician of a Dutch penal colony in Java, Dr. Christian Eijkman. He noticed that hens feeding exclusively on polished rice—the staple diet of the natives—came down with a strange ailment. They were overcome by lassitude which progressed to complete paralysis followed soon by death. Eijkman was able to revive the moribund birds by feeding them the polishings from the rice.

This was a profound discovery. In the first place, here was an unequivocal demonstration that withholding a part of a food from the diet can induce a disease and restoring the same part can cure that disease. Furthermore, having an experimental animal in which a disease can be induced at will is always a great asset. (Our inability to induce pernicious anemia in experimental animals had been a tight brake on our progress against this disease.) When we have convenient, susceptible, experimental animals, forays can be made against a disease from many sides: the internal changes at various stages of the disease can be studied; a variety of possibly dangerous medications can be tried out; once a medication is won, it can be standardized.

Eijkman realized that there is a substance in the outer coats of rice which is essential for health. However, in his interpretation of his findings he went astray. Because the imprint of Pasteur's ideas and personality was so strong, every disease was attributed to pathogenic microorganisms. Eijkman thought there must be something in the rice polishings which "neutralized" the "germs" that cause beriberi.

As often happens, enough scattered information was at hand to weave a pattern of truth. Sir Frederick Gowland Hopkins, professor of biochemistry at Cambridge University, stated in 1906 that animals require in addition to the known components of their diet some unknown "minimal qualitative factors." Six years later Hopkins presented unequivocal evidence for his hypothesis. He placed a group of young rats on a diet consisting of all the known components of milk: protein, fats, sugars, and salts. The weight of the rats on such an artificial diet remained stationary for twenty days. However, if he supplemented the minimal diet with 2 cubic centimeters of milk per day per rat their weight almost doubled during the same period. This was a beautifully designed, clear-cut experiment which could be repeated by anyone, anywhere in the world. The blueprint for the method of search for "vitamines" was drawn up. The catchy name was coined by Casimir Funk, one of the outstanding early workers in this field, who secured evidence that the anti-beriberi substance in rice polishings belongs to a class of organic compounds called amines. In the next three decades new vitamins, a whole alphabet of them, were discovered, but their specific task in the cell remained unknown.

At the time when Hopkins suggested the existence of "minimal qualitative factors" an important parallel discovery was made in England. The intimate relation between the two discoveries did not become apparent for thirty years. The new discovery was destined eventually to throw some light on the role of the vitamins and of bios.

The Buchner brothers were able to ferment sugar to alcohol and carbon dioxide, with cell-free enzyme extracts of yeast. (That was in 1897, forty years after Pasteur had established the nature of fermentation. Interestingly, the forty-year delay was probably Pasteur's fault. He stated that fermentation was the result of the living process of intact yeast cells. No one would design an experiment to challenge the views of so prestigious a person and so formidable a polemicist.) In 1906 two Englishmen, Sir Arthur Harden and W.J. Young, performed a challenging experiment. They took an active yeast-enzyme solution and passed it through a gelatin filter which was known to hold back very large molecules but to allow smaller molecules to go through. Neither the fraction that remained behind on the filter nor the fraction that passed through was able to ferment sugar. But, if the two fractions were pooled, the solution was as good as before in fermenting sugar solutions. For successful enzyme action, then, two factors are needed: large molecules—and we now know that these are proteins—and some smaller molecules to assist the enzymes. These were named coenzymes. But what are coenzymes? Chemists began the tedious task of concentrating solutions of coenzymes with the hope of eventually isolating them. At the same time, other chemists were working on the isolation of vitamins, for example, vitamin B' from rice polishings.

Before the result of these parallel searches is stated, tribute should be paid to a patient Austrian chemist who helped tremendously every other chemist who has to work with minute amounts of substances.

The year was 1910. Fritz Pregl, a professor of chemistry in Graz, was investigating the constituents of bile. Patiently he isolated a couple of hundred milligrams (28,000 milligrams make an ounce) of pure crystalline material of unknown composition. The first step he had to take to establish the composition of his precious substance was to determine the amount of carbon and hydrogen it contained. The available methods for this task required the burning of from 300 to 500 milligrams of material.

The destruction of 300 to 500 milligrams of precious substance which took years of labor to accumulate would provide but one bit of information. Nothing would be left for the dozens of other determinations and manipulations that had to be performed before the complete structure of an unknown substance could be pieced together. Pregl rebelled. He spent the rest of his life perfecting and refining methods so that determinations could be done on one or two milligrams of material. He was spectacularly successful. After the First World War, chemists came to him from all over the world to learn his methods. These disciples, in turn, spread the gospel of microchemistry, for that was the name given to this new technique. Pregl was awarded a well-deserved Nobel Prize for his work. Biochemists were now equipped with sufficiently refined tools to go digging for substances which, like vitamins, were present in their natural sources in minute amounts.[1]

We are now ready to return to the last round in the bios controversy. In 1919 a young American biochemist, Roger J. Williams, published his doctoral thesis, in which he stated that yeasts need "growth promoting substances" in addition to sugar, salt, and ammonia and these substances were identical with "the substances which in animal nutrition prevent beri-beri." Yeasts and beasts need the same vitamin.

At the same time, Williams's older brother, Robert R. Williams, a chemist at Bell Telephone Laboratories, was struggling alone, at his own expense, with the isolation of this very vitamin B˙ from rice polishings.

The next development in the history of bios was the announcement in 1936, by the German chemists F. Kögl and B. Tönnis, of the isolation of bios in pure crystalline form. They obtained, after working up 500 pounds of dried egg yolk imported from China, 1.1 milligrams of a pure substance which they named biotin, in honor of the bios problem. Their method of isolation was the usual painstaking physical and intellectual labor, stretching over several years: subjecting the dried yolk, a rich source of bios, to a variety of chemical separations; testing each fraction for its yeast-growth-

1. Still further refinements of Pregl's techniques enabled our chemists on the atomic bomb projects to master the chemistry of plutonium, from a few *thousandths of a milligram* of it. On the basis of that knowledge the vast plants for its large-scale production were built. We can now analyze with ease a millionth of a milligram or a billionth of a gram.

promoting potency; discarding the inactive, concentrating the active fractions more and more, until the material was sufficiently purified to reward their labors by crystallizing in pure form.

Once biotin became available in pure form its structure was determined and was soon synthesized by Dr. Vincent du Vigneaud of Cornell University. Studies with the new vitamin helped to elucidate several apparently unrelated nutritional problems. Before biotin was isolated a number of researchers reported the existence of several unknown factors that behaved like vitamins. The most interesting of these is the one called vitamin H. If rats are fed raw, uncooked, egg white as the source of their protein they do not thrive. At first a skin rash appears; then they lose their hair; then they become paralyzed, and, if the diet is kept up, they die. They can be saved by cooking the raw egg white, by replacing it with another protein, or by feeding to them along with the egg white either egg yolk or beef liver.

The circumstances pointed to a deficiency disease, and a search was started for vitamin H. (The earlier letters of the alphabet had already been preempted.) At about the same time it was found that a certain microorganism whose habitat is the root of legumes needs an unknown substance in its diet or it perishes. The substance was named coenzyme R, and the search for the substance to fit the name was started. There was still another observation: certain diphtheria bacilli need a well-known compound called pimelic acid for *their* growth. After biotin was isolated and made available, it was found that vitamin H was biotin, coenzyme R was biotin, and the diphtheria bacilli used pimelic acid for the synthesis of homemade biotin.

The disease caused by raw egg white turned out to be due to the deficiency of biotin. There is a substance called avidin[2] in the raw egg white which seizes the biotin and forms with it a tightly knit combination. The biotin cannot then be absorbed from the intestine. (Avidin loses this property when it is cooked.) The disease—

2. A contraction of avid albumin.

egg-white injury—can be produced in humans, too. Volunteers who ate a diet in which raw egg white provided 30 percent of the calories developed the characteristic skin disease in two to three weeks. "This symptom disappeared, but in the fifth week one of the group developed a mild depression which progressed to an extreme lassitude and hallucination. Two others became slightly panicky. The only striking observation in the seventh and eighth week was a marked pallor of the skin." In the ninth and tenth weeks the skin rash reappeared. The subsequent symptoms are not known, for the experiment was halted and the subjects revived by adequate doses of biotin.

The depression and hallucination of one of the volunteers is very significant. This is not an isolated case of the appearance of such symptoms as a result of vitamin deficiency. Volunteers existing on diets deficient in vitamin B' showed similar symptoms, and the dementia of pellagra is well known.

Biotin, a vitamin which is essential to the health of yeasts and was discovered through research on yeasts, is apparently essential for the health of humans as well.

Biotin is effective in very low concentrations. It is one of the most potent biological substances known. A rat needs only .03 micrograms a day. One teaspoonful of the crystals[3] would be enough to supply the daily needs of 1,000,000 rats for 100 days. Since generally in the case of drugs there is a rough relationship between the dosage of the drug and the total weight of the recipient, an approximation for humans can be made, too. A man weighing 150 pounds is about 500 times as heavy as a rat. Therefore, the approximate daily need of biotin for a man is 500 times .03, or 15 micrograms. The teaspoonful of biotin would suffice for 2,000 men for 100 days.

Biotin resolved three different nutritional problems but failed to resolve decisively the bios problem. The fact is that biotin is not bios. Something else is the bios that Wildiers described. Since bio-

---

3. About 3 grams or 3,000 milligrams or 3,000,000 micrograms.

tin was isolated great strides have been made, both in the chemistry of the B vitamins and in the study of the dietary requirement of yeast. There are no less than twelve different members of the vitamin B family. Yeasts need five of these. The other seven they either do not need or they can make themselves. The five needed by the yeasts are the following: thiamin (B_1), inositol, biotin, pyridoxine (B_6), and pantothenic acid. Of these the last one, discovered by Roger Williams, fits the description of the original bios best. For example, the bios described by Wildiers had to be a very sturdy compound to withstand the prolonged heating customarily used in those days for sterilization. Biotin would surely have been destroyed, but not pantothenic acid. Of course, all of this is merely of historical interest, and, as in the affairs of men so in the affairs of yeasts, historical problems are not easily resolved. We do not know with certainty what strains of yeasts Wildiers used, or whether they were pure, homogeneous strains. There are hundreds of different yeast strains, and their dietary needs vary from strain to strain. Furthermore, we also know today that their dietary requirement varies with the time they are allowed to incubate. Given enough time they can make some of these vitamins themselves, among them biotin, but not pantothenic acid. But what name has been given to which vitamin is, after all, of no consequence. What matters is that an old problem has been solved, and during the course of its solution we acquired a great deal of new knowledge. And knowledge is our most valuable possession.

Knowledge not only is rewarding in itself but also it often leads to new developments of practical importance. Fifty years ago the "bios problem" must have appeared ludicrous. Who but "impractical" professors would care whether yeasts need bios in their diet? But from the exploration of the dietary needs of yeasts and of other microorganisms, we learned of the existence of six new members of the vitamin B family. These vitamins improve our own nutrition, one of them is a potent drug against pernicious anemia, and finally, the ultimate in justification, millions of dollars have been made on them. The professors, alas, were not included in the last activity.

How does a vitamin function in a cell? The gross symptoms from the absence of a vitamin are obvious: animals become sick, yeast cannot grow. But what does the vitamin do within the cell to render deprivation of it so devastating? The answer came from biochemists whose interests were enzymes and coenzymes. In the decades following the discovery of the coenzyme of yeast fermentation by Harden and Young, chemists were busily gathering information. It was found that the coenzymes of fermentation by yeast were present not only in yeast, but also in such varied materials as milk, animal organs, and blood. Whether the coenzyme preparation was made from extracts of yeast or extracts of frog muscle, it appeared to be identical. Yeasts and beasts need the same vitamins; yeasts and beasts have the same coenzymes.

These developments provide additional support for Darwin's theory of evolutionary ascent from some common origin. For here is biochemical evidence for the most intimate similarity between yeast and frog; they need the same vitamins; they make the same enzymes; they need the same coenzymes. It must not be inferred that frogs have evolved from present-day yeasts. The implication is that they both evolved from some common ancestral cell in which these basic enzymes and coenzymes were already present.

The coenzyme of yeast fermentation was finally isolated in pure form in 1935. A component of it proved to be niacin, which just about then was proved to be the member of the vitamin B group whose absence from the diet of animals produced pellagra. Niacin was not the only vitamin which turned out to be a component of a coenzyme. As a result mostly of the efforts of Robert Williams, vitamin B_1 was isolated in pure crystalline form and was made available for study. It turned out to be the cornerstone in the structure of a coenzyme which aids in the metabolism of pyruvic acid.

Pantothenic acid, the component of the vitamin B complex which is most likely the historical bios, is also a part of a coenzyme—coenzyme A—which plays a pivotal role in the metabolism of fats.

Thus the real function of vitamins emerges: man and yeasts need vitamins to shape coenzymes to assist their large variety of enzymes.

How does niacin function as a coenzyme of fermentation? To form the coenzyme the vitamin is incorporated into a complex structure containing, in addition to itself, two molecules of sugar, two molecules of phosphoric acid, and another nitrogen-containing compound, adenine. The name of this complex aggregate is abbreviated as NAD. During the course of fermentation by the yeast, hydrogen atoms of sugar molecules are stripped away from the backbone of carbon atoms by an enzyme assisted by NAD. These hydrogens eventually must be combined with oxygen to form water. The hydrogens have to be transported, ferried as it were, from the sugar to those enzymes which can make the hydrogens combine with oxygen. NAD acts as the ferry for the hydrogens. It has the capacity to enter into transient combination with hydrogen atoms and thus shuttle them from one site to another. It does this in close association with the enzyme it assists. The efficiency of such a combination of enzyme and coenzyme in performing their appointed tasks is almost incredible. They can load and unload hydrogen atoms thousands of times in a minute. (The all-time speed record is held by catalase, an enzyme in our liver, a molecule of which can seize and decompose a million hydrogen peroxide molecules per minute.)

Since this book attempts to be a chronicle of ideas, not a purveyor of prescriptions, it will avoid admonition and advice on the choice of appropriate foods for obtaining the full quota of vitamins. Furthermore, we are deluged by information about vitamins these days. To cite a few of these fragments would be pointless; to do a thorough job is impossible. A book called *Vitamins in Clinical*

Practice contains a thousand large pages. Moreover, there is not enough space here to relate the history of the other vitamins, although they are just as interesting as that of biotin.

There are some aspects of our recent knowledge about vitamins, however, which have not yet been incorporated into radio commercials and which throw a light on some of the questions we have been asking.

Why do not animals, for example cattle feeding on a diet lacking in vitamins, develop the symptoms of, say, pellagra? Also, what is the reason for the large individual differences in vitamin requirements which are known to exist in different persons? The answer to both of these questions is the same: the production of vitamins by the microorganisms in the alimentary tract. The alimentary tract teems with microorganisms, most of them harmless, non-pathogenic. Not only are they harmless but, indeed, they are absolutely essential for the life of the cattle on their natural diet. The alimentary canal of the newborn calf (or child) is completely sterile. If the young calf's stomach and gut were to remain forever sterile the animal would have to be restricted to a constant diet of milk or it would perish. However, with its very first meal it acquires the founding fathers of an enormous colony of what is euphemistically called its intestinal flora. The cow, even though it has a multiple stomach, does not have the ability to digest the cellulose which makes up the largest part of its diet. The cow lacks the enzymes to split cellulose. It would starve to death with a stomachful of grass if it were not for its alimentary flora. For these microorganisms are able to convert the cellulose to smaller molecules. While doing this they keep for themselves some of the food and some of the sun energy that had been packed into the cellulose. But they "live and let live," and more than enough is left for the cow. This is a marvelous cooperative enterprise. The cow gathers the food and provides warmth for the little creatures. They pay rent with their labor, for they not only help with the cow's digestion but also provide much of its vitamin requirement. Many microorganisms can synthesize most of the vitamins for their personal needs. As they die, their cell's contents ooze out and the vitamins are absorbed by the cow.

Man, too, plays host to huge colonies of microorganisms in his alimentary canal. Many of these tiny creatures repay this kindness with their homemade vitamins. That this source can contribute a considerable portion of a man's vitamin requirement was made evident by the production of vitamin deficiencies in patients as a result of the prolonged feeding of sulfa drugs. The drugs not only killed the bacteria in the patients' tissues but also wiped out their vitamin factories by the indiscriminate destruction of the alimentary flora.

The amount of vitamins an individual must receive from his diet is dictated by two factors: his body's total requirement and the amount of vitamins his intestinal flora will make for him. Microorganisms differ tremendously in their ability to make their own vitamins; some can make almost all of them, others can make none.

There is an interesting sidelight on the possible role of intestinal flora in lengthening man's life span. At the end of the last century Metchnikoff, the Russian physiologist, was impressed by the large number of hale centenarians he found among inhabitants of the Balkan mountains. He found that sour milk was a staple in the diet of these folk, some of whom claimed to be doubling the Biblical three-score-and-ten. Their milk was soured by certain bacilli which convert milk sugar into an acid, lactic acid. Owing to Pasteur's influence, all scientists were very much preoccupied with the role of bacteria in health and disease. Metchnikoff conjectured that the large colonies of these lactic-acid bacteria in the intestinal flora may crowd out pathogenic organisms and thus promote longevity. The drinking of soured milk became a widespread fad. It would be nice to report now, with our newer knowledge of vitamin synthesis by intestinal flora, that there may be sound basis for enriching our intestines with lactic-acid bacteria and that the vitamins from the lactic-acid bacteria prolong the life of the Bulgarian mountaineers. Unfortunately, there is no basis for this. On the contrary, the lactic-acid-producing bacteria are among the least versatile of microorganisms in this respect. Unless their diet includes most of the vitamins these bacteria die.

With our present ignorance of the causes of aging the only prescription we have for longevity is what can be gathered from statistical studies of people who are blessed with it. Eat well, but not too much; relax; avoid infectious diseases; and above all, choose long-lived ancestors, for heredity seems to be the most important factor.

And, finally, who shall be declared the winner in the Pasteur–Liebig controversy? This frankly prejudiced referee votes for Pasteur. Remember, Pasteur's whole thesis was that fermentation is the result of the living activities of yeasts. In this he was utterly correct. Further, he stated that yeast can be grown in a mineral medium; he used sugar, ammonium, and other salts. Today we *can* grow yeast in a completely "mineral medium" of sugar and salts, plus the five vitamins. The vitamins, incidentally, are far more easily made in the laboratory or in the factory than is sugar. What of Pasteur's error in overlooking the vitamins present along with the yeast cells in his "pinhead" seeding? It is said that Japanese artists purposely introduce a minor blemish in their finished paintings, for only God can make the perfect masterpiece. The "pinhead" then, was the insignificant blemish in the work of the man who was called by the great physician Sir William Osler, the "most perfect man who ever entered the Kingdom of Science."

Why was Pasteur that "most perfect man"? What are the attributes of a great scientist and why is he so rare? Pasteur's grandson, Dr. L. Pasteur Vallery-Radot addressed himself to this question:

Can it be surprising that the scientist of genius should be exceptional? What contradictory qualities he must possess! Besides the gift of observation, he must be endowed with imagination, so he must be a poet. He must be always ready to receive the revelation, what we have called insight; to have that readiness, he must not be narrowly specialized, his knowledge must range over widely varied fields. He must discipline himself to assiduous labor (whereas poets more characteristically wander in a dream world). He must confine himself within the bounds of rigorous experiment, requiring him to bridle his imagination. Lastly, he must have a logical mind, able to draw sound inferences and to synthesize facts observed in the course of his experimentation. Qualities so opposed to one another are very rarely united in a single individual.

Indeed they are rare. In biological sciences there has not been another one like Pasteur.

4. Sugars

Starch is a large, often the major, portion of man's diet. Rice, pota-
toes, and flour are cheaper to produce than cheese, eggs, and meat.
Therefore the majority of mankind lives mainly on those three
starch-laden staples. All too many get very little even of them.

Starch is but one of several different substances which the chemist groups together as sugars. The simplest sugars are grape sugar,
known technically as glucose, and fruit sugar or fructose. Both of
these sugar molecules contain six carbon atoms attached in a row,
festooned with six oxygen and twelve hydrogen atoms. They differ
in the architectural pattern of those hydrogens and oxygens.

A molecule of each of these sugars is grafted together by the
sugar-cane plant to form the cane sugar with which we are all familiar. (Fructose is sweeter than cane sugar. This accounts for the
great sweetness of honey; the enzymes of the bee dismember cane
sugar into its components, fructose and glucose. Saccharin, the artificial sweetening agent, is not a sugar at all. It is a synthetic
organic molecule which, by chance, happens to have an impact on
our taste buds which induces the sensation of sweetness. Saccharin
is not metabolized, therefore it has no caloric value.)

Plants can also clip together hundreds of glucose molecules to
form the multi-sugars—starch and cellulose. The grafting together
of the many small glucose units into the huge starch or cellulose
molecule is accomplished in a simple manner: water molecules are
shed, and the sugar molecules fuse at the shorn sites left by the de-
taching of the water.

We can but marvel at the ingenuity with which nature employs such a simple process. The fusion of smaller molecules, whether of sugars, amino acids, or fatty acids, into the appropriate large molecule is always achieved by the elimination of water.

We must remember that life started in water and continues in water. We are indissolubly wedded to water because the surface of the planet we happen to inhabit abounds in that fluid.

Cellulose, which is the main component of the leaves and stems of plants, is useless to us as a food. We cannot absorb these huge molecules from our intestines; nor do we have the enzymes or the enzyme-bearing microorganisms, as do the cow and other ruminants, to break them up into smaller, usable molecules.

In whatever form the sugars are eaten, starch or cane sugar, they are broken down by the juices of the alimentary canal to simple sugars, which are then absorbed into the blood stream. All the absorbed sugars are converted to glucose; that is the only sugar found circulating in the blood. This does not, however, justify the claim of some candy advertising that dextrose (another name for glucose) is the quick energy food. Cane sugar is split in the alimentary canal so rapidly that it does not differ from glucose in its availability for a normal person.

The amount of glucose in the blood is remarkably constant; it increases in diabetes, but otherwise its level is about the same in all average healthy persons.

The body's heat is derived mostly from the "burning" of glucose; cold-blooded animals such as the frog have less sugar in their blood than we do; birds, which are warmer than we, have more. The writer could not find out whether the blood sugar of the shrew has ever been determined. This tiniest of mammals—it is smaller than a mouse—has a body temperature even higher than that of birds. Undoubtedly its blood sugar is higher, too.

If we could not store so vital a substance as glucose in our bodies, we would have to be eating incessantly to maintain a steady supply of it. That would be a precarious existence. Should we fall asleep we would never wake; for lack of its fuel our bodies would grind to a halt. On the other hand, large amounts of free glucose could not be kept in the body either; it is too readily used up. We therefore deposit glucose in a more stable, less reactive form. Scores of molecules of glucose are hooked together to form this stable reservoir called glycogen. This is the animal's version of the plant multi-sugar starch.

Glycogen is stored all over the body: there are depots of it in the liver, in the muscles, in the kidneys. The living organism husbands and distributes its resources well. As we need glucose, enough of the reserve glycogen is mobilized and is broken down into independent glucose molecules to fill the order. If there is any excess glucose in the blood, as after a meal, it is shipped to the glycogen depots. If energy is needed, the glucose molecule is broken down to release the sun's energy originally packed into it by the green plant. The release of energy is performed with an astounding series of enzyme-motivated reactions.

The earliest glimpses of these reactions were obtained, oddly enough, from studies not of animals but of yeasts. Yeast cells, up to a point, utilize sugars for *their* energy exactly as we do. (They are unable to cope with alcohol, which *we* "burn" with ease to carbon dioxide and water.)

The first step in the metabolism of glucose in the yeast cell or in the human cell was discovered by the same chemists who discovered coenzymes—Harden and Young. They found that yeasts starve in the midst of an abundance of glucose unless inorganic phosphate salts are present. But when phosphates are added to yeasts, they thrive on their glucose. Why the need for phosphates?

Harden and Young found that the yeasts, as the first step in fermentation, hang two molecules of phosphate on the first and sixth carbon atoms of the glucose molecule. Later, other chemists were able to cajole out molecules containing three carbons with phosphate still hooked onto them. An example is phosphopyruvic acid, a compound containing three carbon atoms, which upon losing the phosphate becomes pyruvic acid, the substance which accumulates in the blood of patients suffering from beriberi.

These three-carbon-containing fragments of glucose are found in yeast cells and in elephant cells. What happens to these fragments was obscure until rather recently. Their fate was revealed by the work of many biochemists and the insight of one. Sir Hans Adolph Krebs is a biochemist, trained in Germany, who found refuge in England. From isolated bits of information he pieced together the blueprint of what turned out to be the powerhouse of the cell. Pyruvic acid which contains three carbon atoms is not degraded directly to carbon dioxide. First it combines with another component of the cell which contains four carbon atoms. This seven-carbon-containing product loses carbon dioxide and becomes citric acid, which contains six carbon atoms. Citric acid goes through several enzyme-motivated convolutions, loses carbon dioxide, and becomes a five-carbon-containing product. This substance can again lose carbon dioxide and, after some enzyme motivated alterations, become the original four-carbon compound, which now can fuse with another molecule of pyruvic acid and start the cycle all over. This cyclic process has been called the Krebs cycle in honor of the man who pieced it together. It is an extraordinarily ingenious device which bestows several advantages on the cell. In the first place, the release of carbon dioxide and of energy is very gradual. Furthermore, some of the intermediate components of this cyclic system can be used for the manufacture of amino acids. If a five-carbon-containing amino acid is in short supply, the appropriate five-carbon component of the Krebs cycle is spared from further degradation and is shunted to the enzyme which can convert it to the amino acid. Finally, the Krebs cycle can "burn" not only sugars but fats as well, because fragments originating from fats can also be fed into this exquisitely controlled "fire." There may be as many as forty different enzymes cooperating to achieve this cyclic process. All of the enzymes are packed, in close proximity, within tiny compartments of the cell, the mitochondria.

These enzymes acting in patterned unison probably achieve more than the sum of their parts. I am not suggesting a new version of mystic vitalism reduced to the molecular level, but simply that the integration of the functions of a multitude of enzymes may achieve an effect which transcends the sum of the individual components. We know that proteins acquire unique properties by their very size; in turn, a constellation of such huge molecules may have functional attributes undreamed of in our present-day biochemistry. The enzymes within the mitochondrion regulate themselves, take in raw materials, feed out energy and byproducts; in short, they act as a self-contained, self-regulated powerhouse for the cell.

What other enzymic constellations in other parts of the body can do we do not know. What the French philosopher Henri Bergson called the *élan vital* may well be the bubbling product of such molecular constellations.

How is the energy released from the Krebs cycle used by the cell? Before we can be qualified engineers for this most marvelous of machines we must learn the ABC of the energy of chemical reactions.

All forms of energy are interchangeable: heat can be converted into motion; motion can be converted into electricity, which, in turn, can give light or heat again. Such conversions are often very inefficient. The best steam engine loses about half of the energy of the steam as it converts it to motion.

Energy cannot be destroyed; we can not circumvent its complete liberation by using different paths for its release. We can take a pound of coal and burn it in ample air to carbon dioxide. An amount of heat will be liberated. If we burn another pound of coal in a limited supply of air it will form carbon monoxide, but only about one fourth as much heat will be liberated as before. But if we now burn all of this carbon monoxide to carbon dioxide we get the rest of the original amount of heat. Whether we release the energy in one step or in several steps, the over-all amount is the same.

Now let us take inventory of the energy in glucose. Computations of energy are always based on the chemist's unit weight, or molecular weight. In the case of glucose, this is 180 grams. (One molecule of glucose weighs 180 times as much as one atom of hydrogen.) When a green plant makes glucose from carbon dioxide and water it packs energy into it. Into 180 grams of glucose—about six ounces—are packed 700 Calories[1] of energy. If we burn in a stove the six ounces of glucose we will release 700 Calories of heat.

Now, where are these 700 Calories hidden? They are used to form the bonds that hold together the six carbons, twelve hydrogens, and six oxygens of the glucose molecule. There is energy in chemical bonds. Imagine a dozen large springs from a mattress squeezed into a hat box. A good deal of energy had to be expended to squeeze the springs together before the box lid could be safely locked. If the lid is opened, the jumping springs will release the same amount of energy that was used to put them into the box. This is an analogy of sorts for the energy used to lash the atoms of carbon, hydrogen, and oxygen together to form glucose. (This bond energy has nothing to do with the energy within the nucleus of the atom; bond energy is dwarfed by the monstrous energy of the nucleus.)

The amount of energy in each chemical bond is not the same; some bonds have more energy packed into them than others. The cell gets its warmth and its energy for work from the breaking of the bonds of glucose. Phosphates play a stellar role in the storage of the released energy in a form more convenient for the cell. The howling wind has a lot of energy. The farmer's windmill catches some of that energy and at once puts it to work pumping water. But some of that energy is also stored by the charging of batteries. The wind cannot light up the farmer's house, but the battery can. The heat from a crumbling glucose molecule can not, by itself, make our legs move, but the energy in a phosphate bond can. Phosphate bonds are our batteries. They are the stored energy for life's every need.

What is the mechanism of this battery to which we owe our lives? The battery is a molecule—a molecule called adenosine triphosphate—abbreviated ATP. Onto a molecule of ATP are lashed two special phosphate groups. Ten Calories of energy are packed

1. The Calorie is a measure of heat energy. One hundred Calories will heat one liter (about a quart) of ice-cold water to boiling.

into each of those bonds which secure these phosphates to the ATP. These phosphate-cementing Calories are the only form of energy the cell can use for its many tasks.

Fifty phosphate bonds are formed from the energy released by one molecule of glucose. Fifty times ten Calories are stored from the 700 contained in the six ounces of glucose.

The other 200 Calories which are not captured into phosphate bonds keep us warm. But 500 out of 700, or 70 percent, of the Calories are saved for future work. There is a loss of only 30 percent of the total energy in this transformation. The cell is thus a better engine than the best steam engine, which is only 50 percent efficient in such a conversion.

This stored phosphate-bond energy is used in an ingenious manner. Suppose the cell is in need of a substance the assembly of which requires 30 Calories of energy. Three units of ATP are alerted to act as coenzymes in the cell's assembly line; each ATP unit splits off one phosphate unit and they thus deliver the requisite 30 Calories. The three shorn ATP units in turn require replenishment of their lost energy. Glucose is mobilized from a glycogen depot and is degraded. The energy flowing from the crumbling glucose is used as a cement by the three ATP units to reattach their three phosphates. Then the glucose whittling stops; the three ATPs are ready for any new emergency.

How does phosphate-bond energy move our legs? The bones in our legs are moved by the muscles attached to them. These muscles always come in matched pairs. As one muscle contracts, its opposite relaxes; then the other one contracts, and the first one stretches. We move by such a sequence of alternate relaxing and contracting. All the work in this process is done by the contracting muscle; it moves the bone and stretches its opposite muscle. The energy for this work is provided by the phosphate-bond energy of ATP. The energy for the phosphate bond comes from glucose, and this energy in turn comes from the sun. So we are sun machines like the multivaned toy in the optometrist's window—fantastically complex sun machines, but sun machines nonetheless.

Let us lift the hood and take a look at the machinery. Muscle is made up of long fibers composed chiefly of a protein called myosin. These fibers are teeming with ATP. Dr. Albert Szent-Györgyi, a Hungarian biochemist who had already received the Nobel Prize for earlier work which included the isolation of vitamin C, achieved muscular contraction in a test tube. He managed to free the myosin threads of all their ATP. He then added to these threads, now elongated, some ATP. The long threads curled up instantly on contact with the source of energy. Contraction of myosin thread is contraction of a muscle. This demonstration is a milestone in the history of science, for in Szent-Györgyi's words: "Motion is one of the most basic biological phenomena and has always been looked upon as the index of life. Now we could produce it in a test tube with constituents of the cell."

The availability of the stored energy of ATP for movement is useful to the animal in emergencies. Sometimes the animal needs enormous amounts of energy when there is insufficient time to metabolize glucose. After a vigorous sprint to the bus one may pant for minutes before one can settle down to the calm reading of the morning paper. The panting is a forced intake of large amounts of oxygen needed for the burning of glucose to replenish the ATP used up by the exertion of hurrying to the bus.

The stretching of muscle threads lacking ATP explains the hitherto puzzling rigidity (*rigor mortis*), which sets in soon after death. The ATP present in the muscle is slowly decomposed after death, and, since the enzymes of sugar metabolism are forever stalled, the ATP is never reconstituted. Without ATP, the muscle fibers stretch and cast the corpse into the rigidity of death.

ATP has also been shown recently to be the source of energy for the light which some organisms are able to produce. It is also the source of electrical energy in nerve tissues and in the electric organs of animals which can accumulate and discharge electricity. Thus, glucose is the fuel of the cell, but the energy flowing from it is stored in this most versatile of reservoirs, ATP, which can be tapped to supply all the forms of energy a living organism can generate: heat, light, and mechanical, electrical, and chemical energy. With these recent findings the biochemist of the twentieth century has completed the evidence for the mechanistic concept of life which some biologists of the nineteenth century so enthusiastically espoused.

We have not, as yet, mentioned insulin, which, as everyone knows, is essential in sugar metabolism.

How does insulin fit into that elaborate maze? An astonishing amount of work went into the winning of insulin. This work illustrates the effort needed to expose and understand the role of but one small cog in the complex machinery of the cell.

Diabetes is one of man's worst scourges. There are two different diseases which bear the same name, diabetes mellitus and diabetes insipidus. The only similarity between the two diseases is the same distressing symptom, the passing of enormous volumes of urine. The word diabetes actually means a siphon and the qualifying adjectives, mellitus and insipidus are remnants of the days in medical diagnosis when, unaided by chemistry, the hapless physician was forced to differentiate between the two diseases by the taste of the patient's urine, pronouncing the disease either mellitus (sweet) or insipidus (tasteless). Differentiating the two diseases is about all the physician could do until the early 1920s, when insulin was given to the grateful medical profession and, of course, the even more grateful patients.

Most of the early work in this field was done by physiologists, biological scientists whose interest is the function of cells and organs. They very quickly discovered one of the functions of the pancreas, an organ found just under the stomach. The pancreas produces solutions of enzymes which are poured through ducts into the small intestine, where they split proteins, fats, and sugars.

That the pancreas has other functions, too, was suggested as far back as 1686 by a physician named Johann Conrad Brunner, who thought that the pancreas was in some way involved in the utilization of fats and sugars. Two hundred years later there was complete confirmation of this hunch.

In 1889 two physiologists, Oscar Minkowski and Joseph von Mering, removed, under anesthesia, the pancreas of dogs. The dogs survived the operation, but in four to six hours began to show the characteristic symptom of diabetes mellitus; they were voiding sugar in their urine. As much as two ounces of sugar was lost by one dog in a day. At the same time the sugar in the dogs' blood increased. They became bona fide diabetics.[2]

Prior to these operations it had been known that the duct leading from the pancreas to the small intestine could be completely blocked without harming the dogs in any way. Apparently dogs could do without some of the products of the pancreas: the digestive enzymes which are poured through the ducts. The enzymes of the stomach enabled them to hobble along. The pancreas must, therefore, exert its influence on sugar utilization through some medium other than the juices pumped through the ducts into the intestine.

Complete confirmation of this dual role was provided by a brilliant operation of Minkowski. He removed the whole pancreas of dogs but he immediately grafted pieces of their pancreas under their skin. The dogs survived such operations and led a fairly normal, undiabetic life. With the pieces of pancreas inserted under their skin, there was no possibility of any enzymes getting into their intestines. Whatever the pancreas produced must have gone directly into the blood stream of the dogs.

Insulin was the fruit eventually harvested from these early experiments. Millions of diabetics of this and of yet unborn generations owe their painless days to these discoveries.

They were remarkably fortunate discoveries. Minkowski tried to repeat the production of diabetes in other experimental animals. He was unsuccessful with pigs, goats, ducks, and geese. Finally, with the cat, he was able to duplicate his earlier discovery on dogs.

2. The probably apocryphal story is told that it was the caretaker of the dogs who discovered the sugar in their urine. He is supposed to have noticed that swarms of bees followed the depancreatized dogs.

These two are the only experimental animals which develop a positive, unequivocal picture of diabetes on the removal of the pancreas.

The next great stride forward was made by a young student, Paul Langerhans, who was studying for his doctoral dissertation the structure of the pancreas. He cut thin sections of the organ and saw with his microscope two entirely different types of cells. There were grapelike bunches of cells, and among these there were small islands of cells which were different in appearance.

As early as 1893 it was suggested that it is the island cells—the islands of Langerhans as they came to be called—that produce something which is essential for the normal handling of sugars. There was confirmation of the relation between the island cells and diabetes from the examination of the pancreas of dead human diabetics. Invariably such post-mortem examinations revealed unnatural-looking, degenerated island cells.

There were immediate attempts to apply the newly found relationship between the island cells and diabetes to a possible cure of the dread disease. The spur was its wide prevalence—it afflicts about one percent of our population—and its relentless course through emaciation, muscular weakness, impaired wound healing, and final infection, to death in a few years.

The feeding of organs of healthy animals to patients with diseased organs was an ancient art and superstition. Minkowski himself was the first to try the feeding of the healthy pancreas of other animals to his depancreatized dogs. He obtained no improvement whatever. But the hope that some active principle might be extracted from a normal pancreas spurred on workers for the next thirty years.

This period was by no means fruitless. Chemists developed accurate methods to assay the sugar in the blood of animals in samples as small as a drop. Scores of physiologists plugged away at the preparation of extracts, all of which, alas, turned out to be toxic or impotent, or both, when injected into de-pancreatized animals.

Many were within a hair's breadth of reaching the solution, but history and the Nobel Prize Committee remember only the one who closes that final small gap. Frederick G. Banting of the University of Toronto has been acclaimed as the discoverer of insulin and has received many honors, including knighthood and the Nobel Prize. But we must remember that insulin was not the product of the "flash of genius" of one mind. Scores of scientists from Minkowski on have accumulated new knowledge, and from this pooled information arose a pattern, like a laborious jigsaw puzzle, lacking only the final fragment of information.

The futile question is sometimes asked: What if there had been no Banting? The answer is: Someone else would have extracted insulin successfully somewhat later. There were others who were following the same hypothesis and almost the same procedures as Banting.

This does not imply that there is no genius among experimental scientists. It merely means a different manifestation of genius; in the field of science, genius accomplishes what lesser minds would accomplish later. Creation in the arts is quite different. It is inconceivable that anyone but Shakespeare or Beethoven might have brought forth those very same plays and symphonies. But Newton and Leibnitz independently and almost simultaneously integrated the same mathematical abstractions into differential calculus. The artist extracts his creation almost solely from the riches of his own mind; the scientist evolves in his mind a pattern from phenomena which he and others have pried out from observations of our physical universe. Genius among scientists can be measured in years— the number of years that he is ahead of his contemporaries.

In 1920 Banting, a twenty-nine-year-old Canadian physician, read an article on surgery. It was a description of the effects of the blocking of the ducts leading from the pancreas into the small intestine. The survival of the island cells amidst the degenerating grape cells was emphasized. A brilliant idea took shape in Banting's mind. Heretofore everyone had ground together the whole pancreas in the first step of the preparation of insulin (that name had already been given to the long-sought, active agent of the islands of Langerhans). Was it not possible that the failure to extract an active preparation was due to the destruction of insulin by the enzymes of the grape cells? These grape cells were bursting with potent enzymes which gushed out on grinding. This article in surgery held the answer. Tie off the pancreas; let the grape cells wither and then try to extract insulin from the intact island cells. Lacking the facilities for carrying out this project, Banting applied to the department of physiology of the University of Toronto for help. He was joined by a young physiologist, Charles H. Best, and the two started the quest for the elusive insulin. They tied off the pancreas of several dogs, using standard surgical techniques and care on their patients. After about two months they sacrificed the dogs, removed the pancreas from each, froze these organs, and ground them up in a salt solution which simulates the salt content of the blood.

This mash was filtered free of insoluble debris and the clear solution was injected into a vein of another dog whose pancreas had been previously removed and which by then was in the advanced stages of diabetes. After the injections of these crude preparations the blood sugar of the depancreatized dog was lowered. Insulin was born!

The extraction and purification of insulin could proceed on a large scale once it was demonstrated that the enzymes of the grape cells were the enemies to be thwarted. These enzymes were known to be inhibited by acid. Therefore the whole pancreas glands of cattle were extracted in acid solutions and the extracts were purified somewhat to make them nontoxic. Such an extract was first used on a fourteen-year-old boy in the Toronto General Hospital, who was suffering from severe, hopeless diabetes. His blood sugar was immediately reduced by 25 percent. He was the first of millions to benefit from this new weapon in our all too meager arsenal against disease.

What is insulin and what does it do in the cell? We know the answer to the first question; to the second there is as yet no answer. Insulin was obtained in pure crystalline form and it turned out to be a protein. This explains why feeding whole pancreas or insulin to patients is useless; also why it was impossible to extract it until the enzymes of the grape cells were thwarted. In both cases insulin is destroyed by the protein-splitting enzymes—in the first case by the enzymes in the patient's stomach, in the second by the same enzymes oozing out of the grape cells.

Insulin is poured directly into the blood stream by the island cells. It is but one of a number of substances produced by ductless cell clumps, or glands, which regulate the activity of other cells. The fruits of these ductless glands are called hormones—a word derived from the Greek verb meaning "to rouse to activity."

The preparation of insulin has become a major industry. The pancreas glands of cattle are shipped from the slaughterhouses to pharmaceutical plants for processing. That is why a dish of sweetbreads will most likely be not sweetbreads, the culinary name for pancreas, but thymus, another gland from which the physiologist and the biochemist have not as yet been able to extract anything of greater value.

The potency of insulin preparations was standardized by the health organization of the League of Nations. The signal success of the international standardization of many drugs and vitamins by this organization contrasts starkly with its failures in the political field.

The exact, complete role of insulin in the utilization of sugar is not yet known. It is surprising what a small cog insulin is in the complex machinery of sugar metabolism, in which dozens of different enzymes have already been identified. But the cell's machinery is adjusted with exquisite delicacy. The slightest imbalance at any one point may pile up sugar and flood the diabetic patient with it, wreaking, like all floods, havoc in its path.

How devastating the malfunctioning of just one enzyme can be has been recently demonstrated by studies of another disease involving the faulty metabolism of sugars. Galactosemia is a rare hereditary disease whose symptoms become apparent within a few days after birth. The infant loses weight, vomits, becomes desiccated; his liver enlarges, and, in severe cases, unless the disease is recognized, the patient is lost. The afflicted cannot metabolize the sugar galactose which makes up one half of milk sugar. The other half is glucose or grape sugar. The only difference in the structure of these two sugars is the geometric pattern of the hydrogen and oxygen atoms which are attached to the fourth carbon atom in each sugar.

A normal infant is born endowed with an enzyme which attaches galactose to a coenzyme. Still another enzyme reshuffles the atoms of galactose in this combined form, converting it to glucose. Very rarely an infant is born who, through some monstrous blunder, is unable either to make the first, or attaching, enzyme or to make enough of it. One enzyme which merely helps to refurbish the atoms around one carbon atom can make the difference between life and death of an infant. However, fortunately, if diagnosis is made early and milk sugar is excluded from the diet the symptoms caused by the deficiency of this one enzyme disappear and normal growth may be resumed.

It is sobering to speculate on what our knowledge of diabetes or of insulin would be today if Minkowski and Mering and the physiologists and biochemists who followed them had been forbidden to use cats and dogs, or, for that matter, any experimental animals. Such speculation is not idle, for there have always been a small number of people who have banded together into antivivisection societies whose main function is agitation for the proscription of all animal experimentation.

It is one of the strengths and beauties of a free society that little bands of people like this, can bob up and down on its outer fringes unmolested, while the main stream of society flows on majestically, unconcerned by the little bands which are frantically pouring out a variety of pamphlets, calling on the main stream to be diverted or dammed.

Could these antivivisectionists have their way, however, not only would all research cease, but a great many drugs now available could not be used, for, to insure their safety, they must be tested on experimental animals. Close to sixty people were killed some years ago by a new drug preparation which had not been tested on animals. These corpses, robbed of their spark of life by an "Elixir of Sulfanilamide," are testimony to the absolute necessity of testing a new preparation on animals. Sulfanilamide is quite insoluble in water and therefore cannot be dispensed in a convenient aqueous solution. An obscure pharmaceutical manufacturer tried other liquids as a solvent. He hit on ethylene glycol, and blithely sold such solutions of sulfanilamide to be taken by the spoonful. Unfortunately, ethylene glycol is converted in the body to oxalic acid, a potent poison.

Had the preparation been tested on just one dog first, those people would not have gone to their early graves. What human wealth might have been saved by the life of just one dog. Just one dog, compared to the hundreds of thousands destroyed annually at dog pounds, to test the "Elixir of Sulfanilamide." In the year of this disaster, in New York city alone, 55,000 stray dogs and 150,000 cats were destroyed at the pounds by asphyxiation.

5. Isotopic Tracers

That foods are "burned" in our bodies to carbon dioxide has been known since 1789. We are indebted for the knowledge to the great French chemist Antoine Laurent Lavoisier, who founded modern chemistry and, in a limited sense, biochemistry as well. He boldly declared: "La vie est une fonction chimique."

Lavoisier elevated chemistry to a science by eschewing speculation on the nature of chemical reaction in favor of observation and measurement. He was not the first to apply quantitative methods to the study of chemical or biological systems. But he was one of the first to recognize the imperative necessity of controlling the complete environment of a reaction before a measurement can have any meaning. For example, the Flemish chemist and mystic J. B. van Helmont (1577–1644) set out to determine what a willow tree is made of. He planted a small tree in a previously weighed quantity of dry earth, watered the plant for five years and noted at the end of this time that the willow tree gained 164 pounds and the earth lost only 2 ounces. From this leisurely experiment Van Helmont concluded that all the increase in weight in the tree came from water. It is ironic that not only was Van Helmont aware of the existence of gases (he coined the term from the Greek, *chaos*); he was the discoverer of the very gas he neglected to take into account, carbon dioxide.

A view on the nature of combustion long held prior to Lavoisier was also based on experiments with a similar deficiency. G. E. Stahl (1660–1734), who first postulated fermentation to be the product of the vitalistic vibrations of dead biological material, also provided an imaginative explanation for the mechanism of combustion. When a substance burns, said Stahl, "phlogiston" escapes from it. The phlogiston theory suffered a setback when it was shown that some substances *gain* in weight on combustion. A quick recovery was made, however. Phlogiston, it was decided, was a versatile entity possessing either a positive or a negative weight.

Lavoisier laid the phlogiston theory to rest when he showed with well-designed experiments that combustion is the result of the combination of substances with the "salubrious and respirable portion of the atmosphere." After exploring the nature of combustion of inanimate objects he undertook a similar study of "combustion" in living organisms. With astonishing experimental skill and judgment he showed that the process was the same in both the living and nonliving worlds. A guinea pig and a burning candle both produce carbon dioxide. Furthermore, he measured the amount of heat liberated by the candle and the animal and showed that the heat in each case was proportional to the amount of carbon dioxide produced. (He overlooked the heat liberated by the production of water.)

This genius was ordered to trial by the National Convention and was guillotined in 1794. He was denounced by Marat, whose own ambitions as a scientist he had thwarted. Marat had published his own theory on the nature of combustion, which was essentially a rehash of some stale alchemistic concepts. He denied that oxygen has a role in the process. Lavoisier had proved the views of Marat, the would-be scientist, wrong; but Marat, the politician and rabble rouser, unfortunately had the last word. After the revolution began, Lavoisier became a constant target of vituperative attacks in Marat's newspaper. His crime was that he had been a farmer-general of France. Although during his tenure of office this notorious tax-collecting agency had undergone many reforms, Lavoisier was nevertheless accused and convicted of guilt by association with the once corrupt agency. Testimonials to his great service as a scientist to France and to the cause of the Revolution—he was in charge of standardizing weights and measures—were of no avail. It is reputed that Lavoisier requested a two weeks' stay of sentence so that he might finish some experiments on respiration, but the presiding judge, one Pierre Coffinhal, is said to have replied: "The Republic has no need of scientists; justice must follow its course."

For a long time after Lavoisier's death, life was compared to a burning candle. However, this analogy could not be maintained indefinitely, for obviously a living organism, unlike a candle, is not consumed in its own flame. After the combustion engine was invented, the living organism was compared to that. (Man often belittles the grandeur of life's machinery by comparing it to one of his own handiworks. These days it is the vogue to compare the brain to an electronic computer.)

For a long time food was thought to be merely a fuel for the machine of the body. That the quality of the "fuel" was as important as its quantity did not necessarily conflict with the image of the body as a combustion engine.

However, it slowly became apparent that the living machine is a most unusual one: the parts of the machinery themselves seem to be burning as well as the fuel. The American biochemist Otto Folin, who was appointed, in 1906, the first professor of biological chemistry at Harvard, demonstrated that tissues, too, are broken down independently of the breakdown of foods. He found the excretion of certain waste products from our bodies to be unaffected by the amount or kind of food. He attributed to the metabolism of the tissues themselves these constant waste products.

To what extent, if any, the components of the diet interacted with the tissues could not be determined. For, once the food enters the blood stream—after digestion in the alimentary canal—it disappears. It becomes hopelessly intermingled with those many substances similar to it that are already present in the tissues. Until recently, it was impossible to distinguish such molecules, which were already components of the tissues, from molecules only recently absorbed from the digestive tract.

For example, what happens to a pat of butter we eat? Before we can trace the path of that butter in the body we must first become acquainted with a little of the chemistry of butter. Butter is a fat. A typical fat molecule is built from three molecules of fatty acid and one of glycerol (the glycerin of the pharmacist). Fatty acids are made principally of carbon and hydrogen. The carbon atoms are strung together like beads with two hydrogen atoms linked to each carbon bead. The last carbon of the chain has no hydrogens. Instead, two oxygen atoms are attached to it. This aggregate of three atoms is the acidic group. The chains of carbon vary in length; the shortest is acetic acid (present in vinegar), made of two carbons; the next longer one (present in fats), contains four carbon atoms, the next six, and so they proceed with two carbon increments, up to twenty-four. There is always an even number of carbons, never an odd number, in any natural fatty acid. A natural fat such as butter contains a variety of fatty acids, short and long.

By means of their three acidic groups, three fatty acids are attached to a molecule of glycerol. The coupling between fatty acids and glycerol takes place by the shedding of water. The product made by the coupling of three fatty acids and of glycerol is the fat molecule.

If we cook a fat with hot alkali, we dismember the fat molecule and convert it back into its components: the fatty acids and glycerol. The fatty acids combine with the alkali to form a soap. This operation is the basis of the tremendous soap industry. Soapmaking has been practiced essentially the same way for thousands of years. Any available fat—beef tallow or a vegetable oil—has been cooked with whatever alkali was available. The pioneers in the American wilderness leached out the alkaline ashes of burnt wood; today we use lye produced with electrical energy from table salt. Fundamentally, all soaps are the same. They differ in color and odor, and in the amount of air, salt, and water put into them.

Butter is broken down by enzymes, mostly in our small intestine, into fatty acids and glycerol. The enzymes do, with cool efficiency, what the alkali does in the hot cauldron. The enzyme-produced fragments are absorbed from the intestine and disappear. Where they go and what they do was a complete mystery until some forty years ago. Of course we knew that eventually they must be crumbled down to carbon dioxide and water, but nothing more was certain.

Are they disintegrated immediately, or do they linger in the body for a while? If they linger, where are they? Do they go to the liver, or do they go to the blobs of fat depots which form all too readily around our waistlines? Do they fall apart into carbon dioxide and water at once, or do they disintegrate gradually? Do the different fatty acids have the same nutritional value?

These are the questions which challenged the biochemists' imaginations. They appeared to be questions doomed to dangle before us without answers. For once the fragments of the fat molecule are absorbed from the intestine they become hopelessly intermingled with the pool of other fat molecules already present in the body and become indistinguishable from them.

There have been attempts to label a fat, to hang a bell on it, in order to chart its wandering throughout the body. In one such attempt some of the hydrogens in a fat were replaced by entirely different atoms of another element, bromine. The strategy of this experiment was simple. The presence of bromine is very easy to detect and normally the amount of bromine in tissues is very, very small. If after eating the bromine-labeled fat a rat's liver should contain a large amount of bromine, that would be an indication that the fat entered the liver from the intestine.

While the scheme sounds simple and effective on the surface, actually, labeling a fat with bromine yielded no worthwhile information. Such a drastically altered fat, in which bromine atoms had replaced hydrogen, is an unphysiological substance. The enzymes of the body, which are notoriously fastidious in the choice of substances on which they act, may have nothing to do with such unnatural substances. A label which could go unnoticed by enzymes was needed. Such a label for the fat molecule was provided, not by biochemists, but by atomic physicists. With the aid of this label we not only tracked down the fat molecule but also learned so much that we changed our whole concept of the mechanism of the cell.

Since isotopes have been seared into our minds by the heat released over Hiroshima and Nagasaki only a brief summary of them need be presented here.

The atoms which compose an element are not uniform in weight. The vast majority of hydrogen atoms have the same weight. But there is one atom in every 7,000 which weighs twice as much as one of its more abundant lightweight brothers.

In the case of nitrogen most of the atoms are 14 times as heavy as the common hydrogen atom. (Their atomic weight is 14.) But one nitrogen atom in 270 is 15 times as heavy. (Its atomic weight is 15.) The atomic brothers of different weights are called isotopes. They are identical twins in all respects except their weight.

A light hydrogen atom is composed of a nucleus of a single, positively charged speck of matter, a proton. Rotating rapidly around this portion is a still smaller speck—a negatively charged electron. A heavy hydrogen atom also has only one satellite electron, but its nucleus is different: it contains, in addition to a proton, a neutron—a particle almost equal in weight to a proton, but without a charge on it. Since isotopes differ only in their nuclei, which are not involved in ordinary chemical reactions, their chemical behavior is identical. We know that the heavy and light isotopes enter into ordinary chemical reactions exactly the same way. But what about chemical reactions within the cell? Do the ultrafastidious enzymes differentiate between isotopes? There is a way to decide this question. If the enzymes which build up the body tissues discriminate between the isotopes, one or the other of two isotopes should be more concentrated in the cell than in the inorganic world.

Let us examine the nitrogen atoms which are built into the proteins of our tissues. Let us choose an expendable tissue such as hair for our studies. With appropriate chemical manipulations we can obtain billions of atoms of pure nitrogen from a few strands of hair. Where did these atoms of nitrogen come from? Originally they came from the atmosphere. But before they entered our body, the nitrogen atoms had sojourned in the bodies of many different plants and animals and therefore had participated in a multitude of enzyme-motivated reactions.

Eighty percent of the atmosphere is nitrogen containing the two isotopic varieties, nitrogen 14 and nitrogen 15. (Both of these varieties of nitrogen are stable; unlike the uranium isotopes, they are not radioactive.) Plants and animals are unable to use nitrogen gas from the atmosphere directly. They lack the enzymes for this task. Certain bacteria which grow on the roots of legumes have such enzymes to incorporate the nitrogen from the atmosphere into their cells. From such bacteria a clover might absorb the nitrogen atoms destined for the hair of a human being. If the clover is eaten, the nitrogen, now incorporated into amino acids, would pass into an animal. The nitrogen might be returned to the soil either by excretion or upon the death of the animal. From the soil another plant might absorb the nitrogen. (In this so-called "fixed" form, the nitrogen can be utilized by plants without the aid of nitrogen-fixing bacteria.) Thus, the nitrogen atoms which are now incorporated in human hair may have sojourned in the cells of hundreds of different plants and animals; but in all the multitude of chemical reactions inside a vast variety of different cells the nitrogen isotope was neither concentrated nor diluted. For the proportion of heavy isotope in the nitrogen obtained from hair is exactly the same as that of the heavy nitrogen isotope in the atmosphere.

The enzymes of the cell cannot tell the isotopes apart. But the physical chemist, with his instruments, can. If the isotope is an unstable one the task is easy. The radioactivity resulting from the spontaneous disintegration of an unstable isotope is very conveniently measured with the instrument designed to be sensitive to such disintegrations. There is a constant low level of radioactivity in every living cell. However, if we introduce a radioactive isotope of an element into the cell, the radioactivity increases in proportion to the amount of the isotope introduced.

The stable isotopes entail more work for the researcher. Since they do not send off telltale signals, their presence can be known only by measuring the masses of the isotopes. This, unfortunately, is a much more tedious procedure requiring great experimental skill and expensive hardware.

The first scientist to use an isotopic tracer in biology was George Hevesy, a young Hungarian physicist working in England. In 1923 he immersed the roots of bean plants in solutions containing a radioactive isotope of lead. He traced the ascent of the lead into the stem and the leaves of the plant by simply measuring their radioactivity.[1] Of course, since there is no lead, only carbon, hydrogen, and oxygen, in a fat, Hevesy's studies were of no help in the study of the wanderings of fat in the animal. But his success pointed the way to the use of similar "tracers" should they become available.

The labeling of a fat had to wait more than ten years until Harold C. Urey separated the heavy isotope of hydrogen—deuterium—from the light ones.

The pioneers in the use of deuterium in the study of the overall fate of foodstuffs—or their metabolism—were Rudolph Schoenheimer and David Rittenberg of Columbia University. As the isotopes became more readily available other research teams were formed, until today there is hardly a biochemical laboratory which does not use isotopes as a tool.

To carry out an investigation using isotopes was no easy task. First of all the test compound enriched with isotope had to be prepared. A fat was made which contained not an ordinary fatty acid, but a fatty acid in which an abnormally high number of the atoms were replaced by the heavy isotope of hydrogen—deuterium. This deuterium-containing fat was fed to rats kept in suitable cages so that their urine and feces could be easily collected.

1. According to Dr. Hevesy, his first "tracer" experiments were designed for a somewhat less exalted purpose. Since he was suffering from the endemic affliction of young scientists of that period, impecuniousness, he was lodging in a very modest boarding house. He observed that roast beef was invariably followed by hash on the weekly menu. He had strong suspicions that even the meager leavings of the roast beef from the plates of the boarders greeted them again the following day disguised as hash. Therefore he once sprinkled a trace of radioactivity on his leavings from a meal and next day he surreptitiously carried off a sample of hash to the laboratory. His instruments confirmed his suspicion: the hash was radioactive.

To the amazement of every biochemist and physiologist only a small fraction of the deuterium appeared in the excreta of the animals the next day. Apparently the dietary fat is not burnt up immediately in the body.

Where was the newly eaten fat? Where was the deuterium stored?

Rats were killed one, two, three, and four days after the feeding of the original deuterium-labeled fat. The various organs—the liver, the brain, as well as blobs of abdominal fat were separately cooked with alkali. The released fatty acids from each were isolated and the deuterium in them was determined. Most of the deuterium, therefore most of the original fat, was in the fat depots.

The first intelligence gained from our isotopic detection, then, is that most of the fat from the diet enters the fatty depots first. How long does the fat stay in the fatty depots? The longer the rat lived after the original isotopic meal (they were on a normal rat diet after that), the less deuterium remained in its body. In three days one half of the deuterium, therefore one half of the newly acquired fat, had been used up from the fat depots. In other words, the fat which is incorporated into the body after a meal is used up only slowly. But the animals were not losing weight, therefore as the depot fat was used up, it must have been replaced from more recent meals.

This new intelligence revolutionized our whole concept of life's economy. Previously it had been thought that all entering food was immediately burned up. Only the excess over the body's daily requirement was believed to be stored in depots. These depots, in turn, were believed to be inactive reservoirs tapped only on lean days. The living organism had been visualized as a combustion engine, receiving its fuel—the food—and converting it into energy and waste products without any alteration in the structure of the engine.

But the isotopes told a different story. The information obtained from studies of the fats, later confirmed with other foods, proved that the body is in a constant state of flux. Its tissues are being built up and broken down simultaneously. The molecules which compose our bodies today will be gone in a few days and replaced by new ones from our foods.

The body as a combustion engine is approximated by the train on which the Marx brothers were once escaping from one of their dire predicaments. For want of fuel in the coal car they tore up the coaches, feeding the wooden planks into the engine. Had they been repairing the coaches at the same time from fresh supplies of lumber they would have almost simulated the engine of the body.

The late Dr. Schoenheimer offered a regiment as an analogy for a living organism: "A body of this type resembles a living adult organism in more than one respect. Its size fluctuates only within narrow limits, and it has a well-defined, highly organized structure. On the other hand, the individuals of which it is composed are continually changing. Men join up, are transferred from post to post, are promoted or broken, and ultimately leave after varying lengths of service. The incoming and outgoing streams of men are numerically equal, but they differ in composition. The recruits may be likened to the diet; the retirement and death correspond to excretion." He added, however, that this analogy is "necessarily imperfect."

It is impossible to evoke a perfect analogy to a living organism. It is trite but true that the only analogy to a living organism is another living organism.

Another bounty from the study of fats with the aid of isotopes is the solution of an old riddle in the metabolism of fats. It had been known for a long time that animals can convert sugar or starch into fats. An animal receiving but a small amount of fat in its diet along with large amounts of starch produces huge fatty depots far in excess of the total amount of fat eaten. The fattening of cattle on a diet of corn, which is high in starch and low in fat content, is a practical evidence of this transformation. The corn starch is converted by the cow's cellular alchemy into the characteristic firm

streaks of fat in a good cut of beef. A goose must suffer a fatty en-
largement of its liver to produce a *paté de foie gras* to a gourmet's
taste. It is forcibly fed huge amounts of starchy meals; corn is
shoved mercilessly down its unwilling throat. It produces so much
fat that its liver becomes gross and gorged with fat.

To see whether animals can really make all of their fats, an interesting experiment was carried out by a husband and wife biochemical team, George O. and Mildred M. Burr. Rats were kept on a
diet completely devoid of all fats. Such a diet contains about 25
percent of casein, from which all fats are removed with ether,
which dissolves the fat but not the protein. The diet also contains
over 70 percent of cane sugar. Commercial cane sugar, which is so
pure it contains nothing else, is used. The rest of the diet consists
of a salt mixture and all the vitamins.

At first rats do quite well on such a diet. But after a while it gradually becomes apparent that something is wrong; the rats do not put
on their daily quota of weight. In about seventy days the rats are a
sickly looking lot. Their tails are scaly, ridged, and fragile; pieces of
the tails crack off. The hair is full of dandruff and falls out in
clumps. That their internal organs are also damaged is obvious
from their bloody urine; the kidneys must be cracking too. If the
rats' diet is not changed, these kidney lesions will kill the animals.
But if before the rats reach a moribund state a few drops of fat—
lard or a vegetable oil—are given to them daily, they miraculously
recover.

What is there in a fat which protects the rats against this loath-
some disease? Besides their varying lengths, fatty acids differ in
another respect. They do not all contain their full complement of
hydrogen atoms. In a normal, or saturated, fatty acid there are two
hydrogens for each carbon atom. However, in some fatty acids
there are carbon atoms with only one hydrogen allocated to them.
These are the so-called unsaturated fatty acids. Usually a fat contains a mixture of saturated and unsaturated fatty acids, the liquid
fats being richer in the unsaturated fatty acids.

The various components of a fat were fed to the rats made sick by the lack of fats in their diet. The charm which protected them was not glycerol; it was not the saturated fatty acids; it was an unsaturated fatty acid called linoleic acid.

If minute amounts of this doubly unsaturated fatty acid are fed along with the fat-free diet, the rats lead a normal healthy existence. Why must rats receive this fatty acid and not the others in their diet? Isotopes wielded by Doctors Rittenberg and Schoenheimer gave the answer.

A group of mice on a normal diet received injections of water in which the hydrogen atoms were replaced by their heavy twins deuterium atoms.[2] (Such water has a greater density than ordinary water and is therefore called heavy water. It looks and tastes like ordinary water; only the sensitive instruments of the physical chemist can tell the two apart.) A few days after the injection the mice were killed and their various fatty acids separated.

The saturated fatty acids contained large amounts of deuterium. How did deuterium get into these fatty acids? It could be there only if the mice made these fatty acids, using hydrogen from their body water to string onto the carbon skeleton. After the injection of the heavy water their body water contained not ordinary water but deuterium-enriched heavy water. Since the enzymes do not discriminate between the two isotopes, both ordinary hydrogen and deuterium were strung onto the carbon skeleton of the saturated fatty acids.

The charm, linoleic acid, which wards off the disease produced by the fat-free diet, was also isolated from these mice. It contained no deuterium at all. This is absolute proof that the mice could not make this compound; if they could, there would have been deuterium in it.

The isotopes taught us a new profound lesson. The animal's cells need linoleic acid, but cannot make it because they lack the enzymes for the fabrication of that kind of unsaturation. In turn linoleic acid is needed as a building block for a very potent group of

2. Heavy water was rare and expensive. Only one fifth as much heavy water is needed for an experiment with a mouse as with a rat.

compounds, the prostaglandins which regulate the flow of blood to particular organs.

The animal body must have a huge variety of different substances in its cells for their healthy, smooth functioning. The cells can make most of these. Those they cannot make must come from their diet or they perish. Hence the need for vitamins and, as we shall see later, for essential amino acids and a few other miscellaneous substances such as linoleic acid. (Some biochemists consider linoleic acid a vitamin.)

Some species of animals are not slaves to all of these essential substances. The rat, for example, never comes down with scurvy. It needs no vitamin C from its diet; it can make its own. Plants can make all of their amino acids and vitamins from salts, water, and carbon dioxide. Animals are, in a way, parasites living on the plants. Some microorganisms are parasites, too. We differ only in the degree of parasitism. The yeast can do very nicely on sugar and five members of the vitamin B family. The red bread mold, *Neurospora*, needs only sugar and one vitamin, biotin.

How did animals become so nutritionally dependent? How did they forget the know-how of vitamin making? *Neurospora*, the red bread mold, answered these questions for us. The story of *Neurospora* and what has been learned from it will be related in chapter 10.

While animals have qualitative shortcomings in their manufacturing abilities, work with isotopes revealed that prodigious synthesizing activities take place in animals as well as in plants. Plants are able to fashion a large variety of complex products from carbon dioxide—a molecule containing but one carbon atom. Using the energy of the sun, plants are able to lash together several molecules of carbon dioxide to form elaborate structures. Animals lack the ability to use the energy of the sun directly. But with the stored energy of the sun in the form of carbohydrates, or more specifically of ATP, animals, too, can synthesize complex molecules from smaller carbon fragments including carbon dioxide. For example, we can fashion this way the complex building blocks of our nucleic acids, and a still more complex substance, cholesterol.

Cholesterol contains twenty-seven atoms of carbon, forty-six atoms of hydrogen, and one of oxygen, and is made from several molecules of the two-carbon-containing fatty acid, acetic acid. (If acetic acid which contains isotopic carbon is fed to an animal its cholesterol will contain large amounts of the isotopic carbon.) This complex molecule is present in every cell of the animal body; it is particularly abundant in nerve tissue. In the degenerative disease arteriosclerosis, there is a rise in the cholesterol content of the blood and its deposition in the walls of the blood vessels causes their "hardening." The vessels lose their elasticity. Although cholesterol had been known since 1775, its chemical structure was decoded only about forty years ago. It was found that the kernel of cholesterol structure appears in a variety of hormones—the sex hormones and the hormones of the adrenal cortex, including cortisone. It had been conjectured that one of the functions of cholesterol might be to serve as the raw material for the production of some of these hormones by the appropriate glands. That the conjecture was valid was proved by a biochemist who wielded his isotopic tools with ingenuity. Dr. Konrad Bloch prepared, by chemical methods, cholesterol with large amounts of heavy hydrogen in the molecule. If, after feeding such labeled cholesterol, some of the hormones of an experimental animal would contain heavy hydrogen it would clearly demonstrate that the precursor of the hormone was cholesterol. The choice of the appropriate experimental animal was crucial for the success of the experiment. Since they are so potent, hormones are made only in minute quantities in the animal glands. Feeding the labeled cholesterol to a small experimental animal and then attempting to isolate a hormone from one of its glands would have been futile: the amount of hormone present is too small for successful chemical manipulation. Feeding the precious cholesterol to a large animal such as a cow or a horse would have been prohibitive. However, it was known that some of the sex hormones are normally excreted through the urine, but only in minute quantities (12,000 quarts of urine yield 10 milligrams of the male sex hormone). But, it was also known that pregnant animals excrete one of the sex hormones—pregnanediol—in somewhat larger quantities. The labeled cholesterol was fed to a pregnant woman and from her urine the pregnanediol was isolated. It contained considerable amounts of heavy hydrogen. Therefore the precursor of that hormone must have been cholesterol. In turn, the precursor of cholesterol is acetic acid. Thus, the acetic acid in a salad dressing eaten by a pregnant woman may be incorporated into a hormone molecule by the virtuoso synthetic processes of the human body.

An interesting application of our knowledge of isotopes in biological material is the dating of ancient remnants of life by a method devised by Dr. Willard Libby. The cosmic radiation which is constantly bombarding us converts a very small but constant amount of nitrogen in our atmosphere into radioactive carbon (C^{14}). This C^{14} is inhaled by plants as carbon dioxide and is thus incorporated into every living organism; C^{14} decomposes at a slow and steady rate so that one half of it disintegrates in 5,900 years. During its decomposition C^{14} gives off beta particles which can be determined by an appropriate instrument. A dead organism or any product derived from it can no longer receive radioactive carbon from the atmosphere and therefore the radioactivity of its carbon must diminish at a steady rate. Therefore any prehistoric artifact whose carbon has one half as much radioactivity as a contemporary sample must be 5,900 years old. (The contemporary levels are those determined prior to the detonation of atom bombs. Since 1945 the C^{14} content of the atmosphere and of living things has been increasing.)

If we wish to determine the age of the Dead Sea Scrolls or of the cave paintings at Lascaux we can do it with ease by burning to carbon dioxide a tiny sample of the cloth wrappings on the scroll or the charcoal which was used as paint by the prehistoric frescoist.

The carbon dioxide is fed into an instrument which measures radioactivity, which unerringly counts off the centuries elapsed since the cloth or the charcoal was part of a living thing.

Isotopes are a wonderful tool of detection. With their aid we can follow the path in the body of a particular hydrogen atom in a particular molecule. Questions which seemed unanswerable in pre-isotope days are being answered routinely. Most of our knowledge of the intermediate metabolism—the overall fate from absorption to excretion—of fats and amino acids we owe to isotopes.

But isotopes won't yield miracles. When the newspapers and magazines became isotope-conscious they began to predict medical millennia just around the corner, produced by isotopes. The dragons of cancer, heart disease, or whatever ails one were being slain by the isotopic swords. While making the public research-conscious is extremely valuable, it is cruel to raise false hopes. Isotopes are a tool, a good tool, but just one of many tools.

A tool by itself has never built anything. The scientists whose minds and hands wield the tools are the architects of medical research. Only the ideas of men and women who can dream them will penetrate the startling complexity of a living cell. Only when we have a clear view of the normal pathways in the labyrinth of the cell will we be able to trace the monstrous blunders which lead into the cul-de-sacs of cancer or arteriosclerosis.

Bold, direct frontal assaults have been made on cancer, using radioactive isotopes as weapons. But we cannot as yet control the range of the weapons; it is shooting friend and foe alike. The sensitivity of cancer cells to high-energy radiation such as X rays and radioactivity is well known. However, normal cells are vulnerable too, and unfortunately the staying power of the normal cell under the impact is only very slightly greater than that of the wildly growing cancer cells. There is always some destruction of normal tissues as well.

The dream is to send packages of radioactivity by special delivery into only the diseased cells. We are back to Ehrlich's old problem: to carry the "magic bullet" to specific cells. This time the bullet has, not arsenic, but radioactivity in its warhead; otherwise the problem has not changed. As the French say, "The more things change, the more they remain the same." We can but hope that there is another Ehrlich, not too far away, who will direct the bullet to its mark before the biological scientists unravel the labyrinth of the cell.

6. Amino Acids and Proteins

"Let them eat cake," was the reputed dietary advice of Marie Antoinette to the undernourished poor of Paris.

Came the revolution, the poor, instead of more food, got more advice: "Let them eat glue." This dietary exhortation came from Cadet de Vaux, a physician, who urged Parisians to make soup out of glue, or gelatin, guaranteeing it to be as nutritious as beef soup. The government issued official proclamations endorsing the new substitute for Marie's cake. The Institute of France and the French Academy of Medicine added their authority, praising the ersatz food and cajoling the starving Parisians to become converts to it.

But the Parisians would have nothing to do with such newfangled nonsense. A political revolution they took in their stride, but a revolution of the stomach, that is a serious matter.

About a century and a half later biochemists proved the wisdom of the adamant Parisians. There were attempts to evaluate the food value of gelatin long before the biochemist brought some order to the chaos of the field of nutrition. The most interesting attempt in this prehistoric era of nutritional science was made by M. Gannal, a manufacturer of glue who boldly resolved in 1832 to test the food value of his product. He noticed that the rats which infested his factory ate the raw materials—the tendons, cartilage, and skin of animals—but snubbed his product, the glue, which was obtained by cooking these animal wastes with water.

Gannal conjectured that perhaps rats were merely fastidious about the taste and odor of glue. He therefore decided to perform his feeding experiments on humans. He is to be commended for not urging the consumption of his product on the poor; he fed it to his wife and three children and he himself joined them in their dreary diet. They ate glue, and, to make it somewhat more palatable, glue and bread, for weeks. The result was disastrous. They had violent headaches and intense nausea, and when it became apparent that their health was rapidly deteriorating, M. Gannal reluctantly called off the experiment. He sadly concluded that his product has no food value, indeed it is harmful.

The only thing more difficult than the introduction of a new, fruitful idea, is the banishing of an old, fruitless one. The feeding of glue, or in its more purified form gelatin, kept cropping up in scientific and medical literature for the next hundred years. Convalescents in hospitals, nursing mothers, and infants were fed gelatin. Fortunately, in most cases the feeding period was brief. Nor was gelatin the sole protein in the diet. The conclusions drawn from these improvised experiments were varied, depending upon the susceptibility of the subjects and of the experimenters to autosuggestion.

The field remained chaotic until biochemists entered it with their zoos of experimental animals. First of all, it was established that proteins are an absolute essential in the diet of rats and of dogs. Caged animals, which could not forage for food, were kept on diets completely devoid of proteins. They rapidly lost weight and, unless they were rescued with meals of proteins, they died.

The next question was: Is it the proteins themselves or their amino acids that are essential to the animals?

A wholesome protein, casein from milk, was cooked with acid until it was whittled down to its amino acids. Rats which had no protein in their diet were fed this mixture of amino acids. The rats thrived. This should not be surprising; after all, what happens to the proteins in the alimentary canal of animals? They are broken down by the enzymes into amino acids. Animals do not absorb whole proteins into their blood streams; they absorb only the amino acids. Whether the protein is crumbled down by acid in the flasks of the chemist or by enzymes in the stomach and intestine of the animal apparently makes no difference.

Progress in science is similar to a duel with the mythical multi-headed Hydra. For every question answered, other new questions crop up. Once the biochemist proved that it is the amino acids, not the whole protein, which is essential, he had to face a brand new and far more difficult question. Are all amino acids of equal value to the animal? The obvious way to resolve this question is to feed all of the amino acids to experimental animals and then keep withdrawing the amino acids from such a diet, one at a time.

While this scheme of eliminating amino acids is simple and obvious it could not be carried out until recently because of technical difficulties. Individual amino acids can only with difficulty be cajoled out of a natural mixture of them. They are so alike in chemical behavior that they manage to hide each other from the chemist who is bent on extracting one of them.

Biochemists therefore turned to some natural proteins which were known to be lacking one or more amino acids. Gelatin was found to be such an incomplete protein. Three amino acids are completely missing from it and two others barely put in an appearance. Gelatin is by no means the only protein so poorly endowed with amino acids. Zein, a protein from corn, and gliadin, a protein from wheat, are both lacking in some amino acids.

Rats cannot live on a diet complete in every other way but containing as the only source of amino acids these impoverished proteins. Their growth is stunted, and unless help comes in time, they die. The help is either a wholesome protein like casein or the original, incomplete protein fortified with the missing amino acids.

The growth curves of rats on such a diet illustrate vividly the need for these amino acids. Weaned, litter-mate rats are used in such experiments to rule out individual variation.

A rat on a complete diet containing casein as the protein grows as shown in Figure 6.1. If in the same nutritious rat meal casein is replaced by zein, the protein from corn, the rat will lose weight at

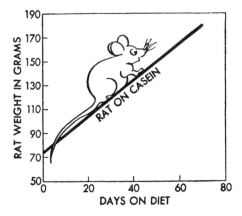

Figure 6.1

an alarming rate, as shown in Figure 6.2. But if we come to the rescue of the moribund rat and throw him a life belt of the two missing amino acids, tryptophan and lysine, the rat will perk up and will begin to gain weight with sufficient speed to walk right off the graph, as in Figure 6.3.

The next question in this logical sequence was: Does the rat find

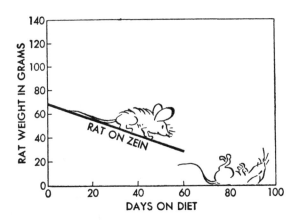

Figure 6.2

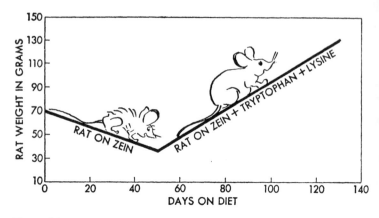

Figure 6.3

every single amino acid essential in its diet? Very little work was required to answer this question. Casein is one of the best proteins; all suckling animals thrive on it. Yet this wholesome protein is low in some amino acids. We must conclude, therefore that all amino acids are not essential, only some of them are.

But which of the twenty amino acids are essential? Had there been twenty different proteins each lacking a different acid, our task would have been easy. Those proteins which could not support growth, we could conclude, lack an essential amino acid. Unfortunately, the various proteins found in nature were not designed for the convenience of the biochemist. There are no such proteins with prefabricated amino acid deficiencies. We had to sort out the essential and nonessential amino acids the hard way.

Biochemists have been nibbling at this problem from the start of the century without making too much headway. About forty years ago William C. Rose, a biochemist at the University of Illinois, started a series of well-designed experiments which were to serve as a powerful beacon illuminating this whole complex field. We learned, under the light of this beacon, that there are no less than ten essential amino acids and that a protein is worth no more than the amount of these amino acids it contains. Dr. Rose decided to keep rats on a collection of pure amino acids rather than on proteins. In this way he could leave out one amino acid at a time and study the effect of the omission on his rats.

It is difficult to convey to the reader what an onerous task this was. Growth studies are long drawn-out affairs. In some cases rats had to be continued on the same diet without a day's interruption for six to seven months. To test twenty different amino acids meant keeping twenty different groups of rats on twenty different diets.

The pedigreed rats[1] used by the biochemist are sensitive creatures; they react to the slightest changes in their environment. Noise, variations in temperature, changes in lighting, the freshness of their food, all affect their growth rate.

Only a few of the amino acids were available commercially, and those were prohibitive in cost. Since these experiments were conducted years before the atom bomb became the exclamation point to the scientist's plea that research pays, money for research was not easily forthcoming.

Rose and his graduate students at Illinois labored long at the accumulation of amino acids. Some they isolated from mixtures of amino acids obtained from proteins; others they made synthetically in the laboratory.

When all the known amino acids were assembled they were fed to rats which were receiving no proteins at all. The rats failed to grow. They could be induced to grow only if the amino acid mixture was supplemented with a bit of casein. Dr. Rose concluded that something else besides the twenty known amino acids must be present in casein and that this unknown substance is also essential for the rats. To track down the unknown factor he began one of those tedious, painstaking searches which the reader by now must recognize as an occupational hazard of the biochemist.

Casein was cooked with acid and the resulting amino acid mixture was put through a variety of chemical separations. Each of these fractions was fed to rats along with their diet of amino acids.

1. The genealogy of these rats is better known than that of the noblest entrant in the Almanach de Gotha.

One fraction when added to the diet of the known amino acids enabled the rats to grow. This fraction, after appropriate purifications, yielded a brand new amino acid, threonine, of whose existence we had not even dreamed, so well had it been hiding among its brethren. Rose started once again to feed his rats all the previously known amino acids plus threonine. The rats thrived.

Then he started a group of rats on a diet containing all the amino acids but one, and he measured their growth. Another group of rats was put on a diet lacking another one of the amino acids and this was repeated until all of the twenty amino acids had their turn at being left out in this most elaborate of musical chairs games.

Ten different groups of rats failed to grow; ten amino acids are essential to the rat. They can do without the other ten amino acids, but if just one of these ten essential amino acids is missing the rat behaves as if it were receiving no protein at all; it cannot grow.

The need for protein, then, is the need for these ten essential amino acids. "By their essential amino acids shall ye value them," should be the injunction to guide us in our choice of proteins. For it has been found more recently, that in the need for the essential amino acids, we humans are akin to the rat. We are not quite as exacting as the rat in our dietary requirement. Our cells can make one of the amino acids and thus we require only nine of them as absolute essentials in our diet.

Now we know why the Gannal family became ill on their diet of gelatin: they lacked some essential amino acids. However, no harm results from eating gelatin. The writer would not recommend gelatin as the sole source of protein in the diet, but as a low-caloried, decorative adjunct to a meal it is useful.

The distribution of the essential amino acids in foods is unfortunate. We find that only the expensive animal proteins—meat, cheese, and eggs—are well endowed with them. The cheaper plant proteins are either low or completely lacking in some of the essential amino acids. Some bean proteins, especially those of the soya bean, approach, but never reach, the animal proteins in their essential amino-acid value.

The impoverished nature of vegetable proteins is interesting in view of the fairly widespread fad for vegetarianism. There is even a Vegetarian political party. The enticing program this party offers is the slaughter of all food animals, and the conversion of all grazing lands into farms to raise vegetables in sufficient abundance to make up for the loss of meat. Whoever concocted this little scheme knew nothing about nutrition and even less about agriculture. The production of meat is by far the most efficient, indeed the only way to utilize over a billion acres of our marginal land. Close to a billion acres of ranch land are not fit for raising anything but grass. This land would be totally unproductive for human nutrition without grazing animals, which alone can convert the grass into meat. Of another half a billion acres of pasture land about three fourths are too hilly for plowing. (We must concede one virtue of the vegetarian platform: it is explicit, a rare quality in this very specialized kind of literary endeavor.)

The practice of absolute vegetarianism from early childhood would be disastrous. The only reason that the effects of the poverty of essential amino acids do not become apparent in vegetarians is simply that there are no absolute vegetarians. A possible exception was the Chinese coolie with his daily bowl of rice, and he is not noted for robust health. Most vegetarians eat milk, cheese, and eggs.

Even the most celebrated vegetarian, the late George Bernard Shaw, had strayed from the true path: he used to take liver pills. He referred to these furtive adjuncts of his vegetarian diet as "those chemical pills."

Partisans of vegetarianism had triumphantly ascribed Mr. Shaw's zestful long life to his partially vegetarian habits. Advocates of health fads indulge all too frequently in this kind of illogic. Well-controlled experiments on the human diet, running for decades on the same individuals, are impossible to achieve. But hope for longevity is fervent, and it is so easy to draw dietary conclusions to nurture that hope.

Those bent on partisan explanations of longevity are particularly apt to commit the "after this, therefore because of this," error of logic. Everything a man has ever done preceded his old age.

The amounts of essential amino acids needed are fortunately so low that a person on an average, well-balanced diet can dismiss all concern about them. For example, the British Ministry of Health advocates the daily consumption of not quite two ounces of a "first-class protein," meaning meat, fish, cheese, or eggs. American authorities advocate about two and a half ounces of mixed proteins per day for an adult. Obviously anyone who eats an egg or two and a fair serving of meat or fish a day is well provided with "first-class protein."

Hypoproteinosis, the condition induced by lack of good proteins, is the most serious nutritional problem in the world today. It is crippling tens of thousands of children in the developing countries. Emaciation, bloated bellies, and either white or red hair on otherwise dark children are the stigmata of protein starvation. The latter, occasional consequences of protein deprivation gave the name, Kwashiorkor, to the disease. It means red-headed boy in some African language. The affluent may also suffer from protein deprivation; it is seen in pregnant or lactating women who are finicky in the choice of their diet.

A multitude of functions in our bodies make the amino acids indispensable to us. In the first place, they are the bricks of which our tissues are built. Eleven of these twenty bricks can be made in our own bodies; that is why these can be left out of the diet with impunity. If one of these eleven amino acids, for example, alanine, is omitted from the diet, we can make enough of it for our needs.

One of the stages in the metabolism of sugars is pyruvic acid. Alanine, the amino acid for whose absence we are about to compensate, and pyruvic acid are very similar in chemical structure. They both consist of a chain of three carbon atoms, two of which— the two end ones—have the same atoms attached to them. They differ only in the atoms that the middle carbons bear. In pyruvic acid there is an oxygen atom attached to the middle carbon, in alanine an atom of nitrogen bearing two hydrogen atoms.

If we happen to need some alanine to build into our body proteins and none of it is forthcoming from the diet, the enzymes in our liver improvise the alanine. They marshal a molecule of ammonia and a molecule of pyruvic acid and clip them together to form alanine. The oxygen atom, which is set free from the pyruvic acid, soon gathers up two hydrogens and they swim away as a water molecule.

Thus we can make an amino acid out of a by-product of sugar metabolism. This is a great asset. We are not slavishly dependent upon our diet for ten of the amino acids. We can produce each of these by making over some fragment from the utilization of our sugars and fats, upholstering such a fragment with ammonia. But the ten essential amino acids that we cannot make must come to us ready made. We can do a bit of an assembly job, we can add ammonia to the appropriate carbon skeleton, but we cannot fabricate the carbon skeleton of the nine essential amino acids. As the mason needs a specially designed keystone when building an arch, so our enzymes need the nine prefabricated essential amino acids for building proteins.

Nor is the assembly into protein molecules the only role of amino acids. They are the most versatile molecules. They are made into hormones and into body pigments, and they are unleashed to disarm poisonous invading substances.

It is beyond the scope of this book to describe the special role each amino acid performs in the body, but one, methionine, will be used as an example. Methionine was chosen for two reasons: it has an interesting history and it performs interesting functions in our cells.

Methionine makes up 1.0 percent of casein and it is essential to rats, men, and microorganisms. Yet we had no idea of its existence until 1922, or of its structure until 1928. It is another one of those biologically important substances which was discovered by work not on animals but on microorganisms. It was discovered by the late Dr. John Howard Mueller, whose chief interest was the dietary requirement of diphtheria bacilli.

This aspect of bacteriology, the nutritional need of microorganisms, was in chaotic disarray at the time. Every science passes through such a muddled period, until coordinating principles are found to weave some pattern out of the mass of apparently unrelated bits of information gathered by the pioneers of the science. Chemistry did not emerge from alchemy until the end of the eighteenth century; bacteriology, a much younger science, started to emerge from its chaotic morass in the second decade of this century. Bacteriologists had tried everything at frantic random to grow bacteria away from living host animals. The prescriptions for raising bacteria, until recently, read like a cookbook for apprentice witches. A concoction might be made of a pig-heart infusion, with a dash of yeast extract, and a soupçon of beef blood. Today, many microorganisms can be grown on completely synthetic diets. As for the others, just give us time, they will eat out of our hands yet—synthetic food.

Mueller was interested in the amino acids needed by diphtheria bacilli for growth. Instead of the bacteriological witches' brew he raised them on acid-cooked casein. He then put the amino-acid mixture through the usual chemical fractionations and tested each fraction as a diet for the bacilli.[2]

From one of these fractions he isolated a hitherto unknown amino acid, methionine, which proved to be essential for the diphtheria bacilli. About ten years later Rose found the same amino acid essential for the rat and only a few years after that the amino acid was being used in the therapy of some diseases of the human liver. The history of methionine offers but one more example of the unforeseen bounties that can accrue from basic research. That not only the public, but even pharmaceutical houses were blind to the value of research is all too clear from Mueller's experience. He

2. This is of course the very process Rose was to repeat about ten years later in his search for the amino acids essential to the rat.

wrote: "The writer recalls somewhat grimly the difficulties encountered in 1920 while attempting to enlist cooperation [of pharmaceutical companies] in getting a hundred pounds of casein hydrolyzed
with sulfuric acid, from which methionine was eventually isolated." The current profit from a day's sale of methionine would
more than cover the cost of the little favor for which Mueller was
pleading.

What does methionine do that makes it indispensable to man,
beasts, and bacilli? Methionine is one of the two commonly occurring[3] amino acids which contain, in addition to the usual elements, the element sulfur. The fate of the sulfur of methionine in
animals has been charted with the aid of isotopes. Methionine has
been made in which the sulfur atom is not an ordinary sulfur but a
radioactive isotope. This sulfur can be traced by the radioactive
messages it sends to a receptive Geiger counter. It turns up in the
other sulfur-containing amino acid, cystine, which makes up almost 15 percent of the proteins of the skin and hair. These are very
special proteins: they are tough and they are insoluble in water. We
should be especially grateful to some creeping ancestor of ours who
first acquired such a skin by a fortuitous mutation. Imagine having
a protein such as egg white for the skin. Getting caught in a rain
would be fatal; our tissues would trickle away and nothing would be
left but our bones, in the midst of a puddle of tissues.

One of the roles of methionine, then, is to provide the raw material sulfur for one of the amino acids of the skin (and of other tissues as well).

Attached to the sulphur atom in methionine is a very simple
group of atoms—one carbon, loaded with three hydrogen atoms—
called the methyl group. It is the simplest pattern in organic chemistry. The methyl group of methionine, too, has been traced with

3. There are more than twenty naturally occurring amino acids. Quite a few
others are found in rarer sources; for example, the octopus yields octopine. While
these rare amino acids apparently play no role in mammalian nutrition, we treat
them with respect. One such amino acid was found by a Japanese investigator in wa-
termellon seeds. A few years later this amino acid was found to be not so rare after
all: it plays a stellar role in the formation of urea in the human liver.

the aid of isotopes. After methionine is eaten, the labeled methyl groups from it turn up in several different substances in the body, substances of vital importance: for example in the hormone adrenalin and in choline, a compound with many roles. Choline aids in the metabolism of fats; it is also part of the impulse-sending mechanism in nerve tissue. It contains three methyl groups in its architecture. These methyls are supplied by methionine.

In the recent past still another pivotal function of methyl groups was discovered by the author and his students, but this will be more appropriately described when we discuss nucleic acids.

A deficiency of methionine and of choline can be disastrous for the animal. Its liver becomes diseased and degenerated, and unless these nutrients are restored, the animal or human patient dies. From the curiosity of a bacteriologist about the dietary needs of the diphtheria bacilli, a medication for humans was harvested.

And what happens to the remnants of the methionine molecule?

We can be sure it is not wasted. The methionine molecule shorn of its acidic group and of the sulfur and methyl group via a complex series of enzyme reactions becomes part of a very important group of substances for the cell, the polyamines. As the name implies these compounds have more than one amino group on them and consequently, with the absence of the negatively charged acidic group they are highly positively charged. They therefore can neutralize some acidic components of the cell such as nucleic acids.

Finally this most versatile of amino acids enters intact into a combination with other amino acids to form that most marvelous of substances: the protein of our tissues. The protein molecule is one of nature's masterpieces of complexity. In the elaborate pattern of that molecule is locked the secret of life. The proteins are the mechanics of life: they fabricate its tissues, regulate its energy, and help assure its perpetuation. How the amino acids are lined up faultlessly to form these molecules with almost magic attributes will be the subject of later chapters.

There is no extensive storage of proteins in our bodies. We can store huge amounts of fats in our fatty depots and we store sugar in the glycogen depots, but we store no proteins in excess of the amounts needed to make up our tissues. On the one hand this is a serious drawback, since we must, therefore, depend on a steady supply of essential amino acids from our diet. But then, many proteins are active enzymes. The unchecked accumulation of highly potent protein molecules could become disastrous: we might grow to monstrous size and an uneven distribution of proteins in our bodies could wreak havoc with the delicately adjusted balance of the various parts of the whole individual. The accumulation of large blobs of fat around our abdomen does damage only to our vanity. The deposition of similar amounts of protein would be lethal. Even hibernating animals, which must lay in a large store of food in their bodies for a long siege of starvation, store only fats, not proteins.

If an animal happens to receive an overgenerous supply of amino acids, in excess of its immediate needs, it metabolizes them into urea, carbon dioxide, and water; or, if it requires no immediate source of energy, it can convert the excess amino acids into sugars and fats, the depots of which can be readily augmented. Earlier in this chapter the process by which an intermediate in sugar metabolism, pyruvic acid, can be converted into the amino acid alanine, was outlined. If, on the other hand, an animal is encumbered with an excess of alanine, it reverses this process: it converts the alanine into pyruvic acid. Certain enzymes—mostly in the liver and kidney—remove the ammonia from the amino acid and replace it with an atom of oxygen. Other enzymes combine the newly formed ammonia with carbon dioxide to form urea which is then excreted. Thus two fairly toxic waste products, ammonia and carbon dioxide, are efficiently eliminated in one step. The molecule of pyruvic acid is either further metabolized to carbon dioxide and water, or it can be converted all the way to glucose. The pyruvic acid can, by a more elaborate process, be converted into a fat, too. Thus the proteins in our diet are ready sources of carbohydrate and fat. It is well to remember this, since the nonfattening nature of proteins has been over emphasized by the designers of reducing diets. While it is true that proteins have lower caloric value per unit weight than fats or carbohydrates, nevertheless, unfortunately for those bent on acquiring a more stylish figure, a thousand calories coming from lamb chops are just as fattening as a thousand calories from rice pudding.

7. Atomic Architecture

The rocket that hurled Russia's first Earth satellite into orbit also served as a starting gun for a marathon debate on American education. Hardly a day passes that we do not read in our newspapers of demands for more sciences in our curricula, or of counterpleas: "Let us not neglect the humanities."

Josiah Willard Gibbs, the great American theoretical scientist of the nineteenth century, once attended a faculty meeting at Yale where a debate droned on about the relative merits of mathematics versus languages in an undergraduate curriculum. Apparently then, as now, nothing aroused members of a faculty to greater heights—and lengths—of eloquence than the carving up of a student's academic carcass. Gibbs, who was retiring to the point of being a recluse, listened to the harangues about the superior merits of languages. Finally he rose and is reputed to have said: "Gentleman, mathematics *is* a language"—and left.

One is tempted after hearing much of the current debate on the humanities versus the sciences to paraphrase Willard Gibbs: "Gentlemen, the sciences *are* humanities." They are but slightly different branches on which creative imagination blooms. Their seed is nurtured by the spirit of the time; their fruits are savored by generations; their impact is sometimes identical.

In 1857 two Frenchmen, a novelist and a chemist, published their major works. Flaubert with his *Madame Bovary* launched the modern novel of realism; Pasteur with his study of fermentation started us on a path of realism in dealing with life's processes. The one cast away the idealized images of human emotions and behavior, the other lifted the miasma of vitalism which befogged our view of biological processes.

Even aesthetically a creation of science can approach man's noblest works of art. It is only unfortunate that the aesthetics of science has a very limited audience. Any one with sight and soul can enjoy a Bernini statue, but only those versed in the language of science can savor the beauty of a scientific creation. Some years ago I overheard an instructor at Cambridge University extolling to a group of undergraduates the beauties of King's Chapel. The teacher felt this was the ultimate in architectural beauty, never to be surpassed or even equaled by another work of man. Neither the instructor nor his students were aware that at that very moment, only a stone's throw away in a poorly equipped laboratory, a young biochemist was performing an architectural feat of such subtlety and beauty that his work, I felt, was more than a match for King's Chapel. The creator of that noble vaulted structure and the biochemist wielded different tools, worked in vastly different dimensions, but each brought to his task the ultimate in the skills and the knowledge of his century and, as genius will, each leaped ahead of his contemporaries to create an enduring triumph of his time, the one in stone, the other in atoms.

To one who understood him, listening to Dr. Frederick Sanger unfold his tale on the decoding of the structure of insulin was just as thrilling as King's Chapel at twilight with the organist weaving a Bach fugue. I shall try to convey to the reader the extraordinary ingenuity with which the problem of the structure of the protein hormone insulin was attacked and the sheer beauty with which it unfolded.

Proteins are composed of long chains of amino acids. The links of amino acids are forged together by the electronic forces of the atoms. To achieve a union, two amino acids together shed a molecule of water. Part of the water molecule comes from the acidic group of one amino acid, part from the amino group of the other. The electronic forces which originally had held onto the atoms which form the water now graft the two amino acids together. In other words, an atom fragment is carved out of each amino acid and fusion takes place at the shorn sites. Such a couplet of amino acids still has the acidic group of the second unengaged. This can reach out to another amino acid and so they continue until vast accretions accumulate. Some proteins are as much as a million times as heavy as a hydrogen atom. The individuality of each protein chain is shaped by its sequence of amino acids. From that sequence is derived the physical and biological attributes characteristic of each protein.

Since there are twenty different amino acids, the variety of protein molecules which can be woven out of them is beyond imagining. Let us look at a simple example.

If only the three amino acids A, B, and C are used to form a triplet—a so-called tripeptide—the following are the different products possible:

$$
\begin{array}{ccc}
A & B & C \\
A & C & B \\
B & A & C \\
B & C & A \\
C & A & B \\
C & B & A \\
\end{array}
$$

The permutations possible can be predicted without listing them. It is factorial 3, symbolized as 3! which is equal to $1 \times 2 \times 3 = 6$.

In longer chains of amino acids the number of possible variations fairly leaps into astronomical magnitude. With twenty different amino acids the permutations possible are 20! which is equal to two billion billions (2×10^{18}). And such a chain would be puny as proteins go, weighing only about 2,000 times as much as one hydrogen atom.

The way in which amino acids are linked together has been known for over fifty years from the work of the great German chemist Emil Fischer. But until thirty years ago it appeared that the decoding of the *sequence* of amino acids in a protein would present unsurmountable difficulties. That such a feat has been accomplished is a tribute to the ingenuity of three English chemists, Dr. A.J.P. Martin, Dr. R.L.M. Synge, and Dr. Frederick Sanger.

Doctors Martin and Synge provided an extraordinarily simple analytical tool for the detection of the presence of each amino acid. Prior to their achievement, the analysis of amino acids was a grueling task with uncertain outcome. Amino acids are so similar in chemical and physical behavior that their separation in pure form—a *sine qua non* of analysis—was tedious and incomplete. For example, thirty-five years ago a well-known chemist completed the first reliable assay of glutamic acid in the protein of milk. It took him a year to complete his study, and he had to start with a hundred grams of the protein. Today a skilled assistant can complete a dozen assays of glutamic acid in twenty-four hours by paper chromatography, and he needs but one millionth as much of the starting material. This technique which did for protein chemistry what the plow did for farming, is based on a very simple principle. It depends on the fact that liquids will rise, defying gravity, along thin fibers or tubes. If a large sheet of filter paper is rolled up and the lower edge of it is inserted into a dish of water the moisture will creep up the paper. If kept in that position for twenty-four hours, the water will rise to a height of from twelve to fifteen inches. Martin and Synge noted that if minute amounts of chemicals, say amino acids, are placed on the bottom of such paper they will migrate upward with the liquid; the relative height to which they travel is an unvarying characteristic of each amino acid. There is a simple reagent which when sprayed on the dried paper will reveal the area to which the amino acid had risen by turning the spot lavender. We thus have a delightfully easy method for the detection and identification of as little as a hundred thousandth of a gram of an amino acid. We can compare the rate of migration of an unknown substance with that of a genuine sample of an amino acid, and if they travel together and give the same color reaction, they are identical. See Figure 7.1.

Doctors Martin and Synge received the Nobel Prize for their

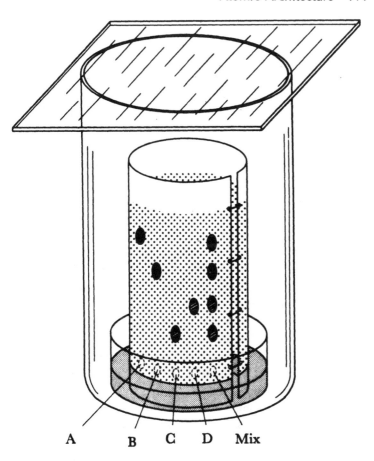

Figure 7.1. A Paper Chromatogram

simple scheme, for they opened up frontiers of biochemistry which had previously remained barred due to clumsier analytical tools. For example, we can extract from fossils minute traces of organic matter and ask them via paper chromatography what they contain. The astonishing answer from these remnants of creatures that lived as long ago as three hundred million years is that their bodies were constructed of the same amino acids as ours.

With this tool in hand Dr. Frederick Sanger boldly ventured on a vast enterprise: to determine the structure, that is, the exact sequence of amino acids, of a protein. He chose insulin for the first assault for three reasons: it is the smallest protein known with definite physiological properties; it was available in pure crystalline form; it was relatively cheap, an important consideration for a chemist working on a minuscule budget. (In Dr. Sanger's laboratory washbowls long discarded from Victorian bedrooms served as makeshift scientific utensils.)

It was known from the probings of several chemists that insulin contains only sixteen of the known amino acids. It was also known that some of these amino acids are repeated as many as six and seven times. The average weight of an amino acid is 100 times that of hydrogen, therefore a chain of sixteen amino acids would weigh only 1,600. However, insulin weighs 6,000. Therefore, some of the amino acids must be repeaters in the structure.

With this information at hand Dr. Sanger was ready to tackle the most intricate cryptogram ever to challenge the human mind. His first goal was to devise a method of pinpointing the first amino acid in the chain. That is the one whose amino group is not forged into a link. There had been previous attempts to achieve such a labeling, but none were too successful. An agent for such a task has to be a molecule which seeks out only a free amino group and which adheres to it so tenaciously that the amino acid chain can be dismembered by heating with acid without removing the label. If in the debris of amino acids the one with the label can be found, we can be certain we have the head of the chain. Dr. Sanger devised a remarkably effective label which meets all the requirements and has an added fillip: it is bright yellow. It thus serves as a flag amidst the amino acids, all of which are white. Once this compound, now called Sanger's reagent, was on hand decoding could start.

Insulin and Sanger's reagent were permitted to unite, and the yellow product was dismembered with strong acid. The solution of amino acids was then subjected to paper chromatography. The sixteen amino acids arrayed themselves as anticipated, but on inspection they revealed a surprise. There were not one but two different amino acids carrying Dr. Sanger's label. Therefore there must be two amino acids—they turned out to be glycine and phenyl-alanine—whose amino groups are not engaged in the weaving of the insulin molecule.

This intelligence could be interpreted only one way: insulin is not a single strand of amino acids; it must be a double one. How are these two strands held together? Insulin is rich in the sulfur-containing amino acid cysteine, which has a unique attribute. It has not two but three hooks with which it can form protein links. In addition to its amino and acidic groups its sulfur atom can reach

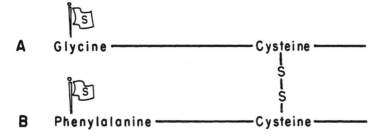

Figure 7.2. Insulin with Sanger's Pennant Attached

out to the sulfur atom of another cysteine in a neighboring strand and forge a fairly strong link.

Our knowledge up to this point is summarized in Figure 7.2 in which a flag designates Sanger's reagent (the exact position of the cysteines is unknown).

The next order of business was to separate the two chains (A and B). This was relatively simple, since the sulfur links were known to be easily broken by oxidation. A mixture of the two chains was thus obtained, but they had to be separated. At this point Dr. Sanger's ingenuity was matched by good luck: the two chains turned out to have different electrical charges on them. Molecules with different charges migrate in different directions in an appropriate electrical field, and we are therefore able to separate such a mixture into two homogeneous components.

The two different chains produced by the cleavage of the sulfur bridges were separated and purified. Chain A, which had glycine as its first amino acid, had about twenty amino acids in it; chain B, whose guide bearer was phenylalanine, had about thirty.

Chain B was tackled first because it promised to be easier. There are a larger variety of amino acids in it, therefore there was a likelihood that some of them might appear only once. In chain A, with a smaller variety of amino acids, the probability of repeated appearance is greater, which places an extra burden on the decoding.

Chain B was treated with Sanger's reagent and was subjected to very mild fragmentation. This is done with weak acid and very gentle heating. Under these conditions the cleavage is incomplete, yielding fragments of the chain containing two, three, and four amino acids still hooked together.

Such a mixture of debris was paper chromatographed and allowed to migrate according to the propensities of the components. (Their rate of travel is slower than that of individual amino acids.) Dr. Sanger observed several different yellow spots. Any labeled phenylalanine which was free of all other amino acids formed one spot. A phenylalanine which still had the next amino acid attached to it formed another spot, one with two amino acids attached to it still another, and so on.

Each of these spots on the paper was cut out, the amino acid chains leached out of them, and each solution evaporated. Dr. Sanger now had a series of fragments of the original chain.

All of the leached-out spots were cooked with acid to break them down completely, and they were analyzed by paper chromatography for the identity of the amino acids.

Dr. Sanger was successful in finding fragments which contained only two amino acids, the labled phenylalanine plus another amino acid, which turned out to be valine. Now the decoding was under way. The first two amino acids of chain B are: phenylalanine-valine. A triplet of amino acids was also found which contained labeled phenylalanine, valine, and a third amino acid, aspartic acid. Therefore, the sequence of the first three amino acids must be

Phenylalanine—Valine—Aspartic Acid
    1   2   3

A quadruplet of amino acids revealed the sequence:

Phenylalanine—Valine—Aspartic Acid—Glutamic Acid
    1   2   3   4

A quintuplet of amino acids yielded the following sequence:

Phenylalanine—Valine—Aspartic Acid—Glutamic Acid—Histidine
  1   2   3   4   5

This was the longest of the chains still containing labeled phenylalanine which Dr. Sanger could find. At this point he resorted to other tactics. He turned to the amino acid complexes which did not have the labeled number one attached to them. In other words, he started examining the mixtures which originated from further along the chain. He found a triplet which contained amino acids 4 and 5 and a new one which was not 3. A chain of thirty amino acids can form only ten triplets. Since there are sixteen different amino acids to be distributed among those ten triplets, the probability of 4 and 5 appearing more than once is very small. Therefore the triplet containing 4 and 5 and the new one most likely was 4—5—6. Amino acid 7 was found by obtaining a triplet containing 5—6 and a new one.

Dr. Sanger continued hunting and matching overlapping fragments with the sensitivity of an artist, the concentrated intensity of a genius, and the ingenuity of a truly original mind.

The sequence of some of the fragments from further along the chain could be decoded only by labeling *their* first amino acids with Sanger's reagent and performing a stepwise fragmentation on them. As the sequence was laboriously wrested from each fragment it was fitted into the larger pattern until, finally, thirty amino acids were unequivocally delineated.

He then tackled chain A and, using the same strategy, decoded a sequence of twenty-one amino acids.

He next attacked the question of the location of the sulfur bridges through which the two chains are connected. It is easier to visualize his approach if we line up the two chains side by side, designating the amino acids by numbers (see Figure 7.3).

In chain A cysteine appears four times: 36—37, 41, 50. In chain B it makes its appearance only two times: 7, 19.

The question is which of the six sulfur atoms are the abutments through which the bridge or bridges are formed. To answer this question Dr. Sanger's ingenuity once again provided a unique approach. He subjected whole insulin, with the sulfur bridges intact, to disintegration by enzymes which were known not to sever sulfur bridges, and he started once again to hunt for fragments. This time he sought only fragments which contained cysteine. He found novel sequences which exist neither in chain A nor B. For example, he found a fragment of four amino acids, as in this diagram:

8	7	37	38
Glycine——Cysteine——Cysteine——Alanine

Since such a sequence appears in neither chain A nor B, the two cysteines must come from the two different chains carrying a neighbor with them; therefore, there must be a bridge between 37 of A and 7 of B.

Similarly, he found another bridge between 50 of A and 19 of B.

Thus, after a decade of intense dedication Dr. Sanger could unfurl one of the proudest pennants signaling a human achievement, the structure of a physiologically active protein (see Figure 7.4).

The sequence of amino acids in insulin reveals no pattern of any kind; it appears to be just one of the trillions of random sequences possible. Yet somewhere in that sequence resides the physiological potency of this hormone. Where, exactly, we do not know. Part of the chain can be altered without impairing the function of the hormone. Dr. Sanger's original work was done on beef insulin. But as the sequence of amino acids was determined in other insulins a surprise was in store for us. Chain B was the same in each one examined, but in chain A the variations shown in Table 7.1 emerged.

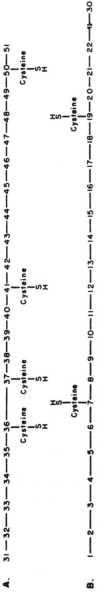

A.

31 —32 —33 —34 —35 —36 —37 —38 —39 —40 —41— 42 —43 —44 —45 —46 —47 —48 —49 —50 —51

Cysteine   Cysteine   Cysteine   Cysteine

B.

1 —2 —3 —4 —5 —6 —7 —8 —9 —10 —11 —12 —13 —14 —15 —16 —17 —18 —19 —20 —21 —22 —1—30

Cysteine   Cysteine

Figure 7.3. Position of Cysteine in the Two Chains of Insulin

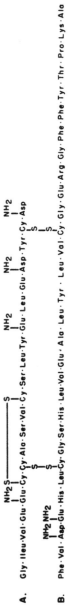

A.

Gly·Ileu·Val·Glu·Glu·Cy·Cy·Ala·Ser·Val·Cy·Ser·Leu·Tyr·Glu·Leu·Glu·Asp·Tyr·Cy·Asp

B.

Phe·Val·Asp·Glu·His·Leu·Cy·Gly·Ser·His·Leu·Val·Glu·Ala·Leu·Tyr·Leu·Val·Cy·Gly·Glu·Arg·Gly·Phe·Phe·Tyr·Thr·Pro·Lys·Ala

Figure 7.4. The Structure of Insulin

These amino acids certainly cannot have unique roles within the structure of insulin. They appear to serve merely as struts to hold the structure to a definite dimension and shape. The variability of amino acids in the insulins of different orders of mammals—indeed between two different families of whales—is interesting from the point of view of evolution. Nature paints the image of life with bold uniform strokes; however, the loose bristles on the edge of her brush do often trace minor patterns distinctive of each species. No mammal has survived a mutational change involving the total loss or even gross changes in the structure of insulin. Even the changes

Table 7.1. Amino Acid Variations in Chain A

Animal	38	39	40
Cattle	alanine	serine	valine
Pig	threonine	serine	isoleucine
Sheep	alanine	glycine	valine
Horse	threonine	glycine	isoleucine
Sperm Whale	threonine	serine	isoleucine
Sei Whale	alanine	serine	threonine

in the three permutating amino acids 38, 39, 40 are relatively minor: the amino acids involved are quite akin in size and other physical attributes. However, all amino acid replacements are not quite so benign. As we shall see in the next chapter in some cases the interchange of one amino acid in the hemoglobin molecule of the human can spell the difference between life and death, literally.

Several different chemical forces determine the geometric form of a protein molecule. In the first place there is the permutation of alignment of amino acids which determines the sequence of side groups that stick out of the backbone of a protein. Since the core of the protein molecule—the repeating links of the chain—is identical in all proteins, the characteristic attributes of different species of proteins must be derived from the ordered array of the side groups. We call this the primary structure of the protein molecule. As we have seen, in the structure of insulin the amino acids along the chain can form liaisons on their own initiative. Two cysteines within easy atomic reach of each other can form a loop of atoms tied together by the two reactive sulfur atoms. Or, cysteines from two different chains can reach out and build a bridge and thus fuse the two strands of amino acids into a more rigid, ladderlike structure. This is called the secondary structure of the protein molecule.

In addition, there are strong chemical forces which orient proteins into three-dimensional shapes. This architectural effect was deduced by one of the most versatile of living American scientists, Linus Pauling. Pauling explained certain regularities revealed by X-ray analysis of proteins as stemming from a repeated helical arrangement of the core of the protein structure. He pictured the atoms which form the links between the amino acids as not stretched out like a limp string but as coiled into long spirals. Forces between atoms in adjacent layers of the spiral provide added strength and rigidity and form the so-called tertiary structure. In turn, the coils of the protein molecule are intertwined to form the final edifice within which may reside the wondrous potency of an enzyme or of a hormone.

Another monument of architectural biochemistry was shaped by Dr. Vincent du Vigneaud of Cornell University, who was our foremost expert in the biochemistry of sulfur compounds, cut his scientific eye teeth, interestingly enough, on the chemistry of insulin. Then, about forty years ago, his attention was attracted to the hormones of the posterior lobe of the pituitary gland, because these hormones, like insulin, were also known to be rich in sulfur compounds.

Since the work on the hormones of this gland is repetitive, let us concentrate on just one hormone. Oxytocin has profound physiological effects on the pregnant human female. Minute amounts of it can induce labor by causing uterine contractions, and it can also induce the flow of milk.

The isolation of the hormone was one of those Herculean feats lasting over decades. It entailed the extraction of tens of thousands of beef glands, the application of new tools of purification, and, finally, the patient cajoling out of the pure product from the last traces of contaminants.

Dr. du Vigneaud analyzed the pure hormone and found that it contained eight amino acids, each of which appeared only once in the molecule. This was fortunate because such a structure is relatively simple. Simple, however, only in comparison to the structures of proteins; for oxytocin, even with only eight amino acids, can exist in 8! or 40,320 different forms.

The structure of the hormone was determined by techniques essentially the same as those described under the work on insulin. The first amino acid was determined with the use of Sanger's reagent, the various amino acid sequences were deduced by the partial fragmentation of the molecule and identification of the amino acid complexes by paper chromatography.

From this pooled information Dr. du Vigneaud decoded the sequence of the eight amino acids. The relatively simple structure of oxytocin offered to Dr. du Vigneaud the tantalizing possibility of its synthesis.

Dr. du Vigneaud came to this final asault, the synthesis of the most complex biological material ever attempted, with unique qualifications. He had previously demonstrated rare genius in structural biochemistry: he had synthesized the vitamin biotin, and he was the first to synthesize penicillin.

It would require a whole book to describe the logic and the techniques of organic chemistry on which the planning and execution of the synthesis of oxytocin was based. Synthetic organic chemistry is one of the most extraordinary achievements of the human mind. It has been built on a curious mixture of rather messy and pedestrian experimental techniques and an ingenious system of logic which uses crude symbolic images to describe a variety of attributes of the compounds of carbon. The images depict both structure and function. The validity of this system of logic has passed untold pragmatic tests: hundreds of products of the living cell have been duplicated in the flasks of the organic chemist. One of the most defiant challenges to this system of inference was the structure and synthesis of oxytocin. Does this complex man-made hormone have any physiological potency on man? Dr. du Vigneaud handed over to clinical associates his product to be compared with natural oxytocin. "They were found to be indistinguishable in effectiveness. Approximately a millionth of a gram of either the synthetic or natural material given intravenously to recently parturient women induced milk ejection in 20 to 30 seconds."

The synthetic oxytocin is now available commercially and is used by obstetricians for the induction of labor in the terminal stages of pregnancy. Not the least valuable asset of this drug is our ability to time with its aid the onset of parturition at the convenience of patient and physician. Thus a product shaped by the hands of the chemist may produce a technological revolution to obstetrics: it will no longer be a nocturnal profession.

The biological activity of the synthetic material proving it to be identical with the natural was the laurel wreath on twenty-two years of inspired effort.

The Nobel Prize Committee bestowed its accolade upon both Dr. Sanger and Dr. du Vigneaud. Their achievement equals, many of us think, the stunning accomplishment of the atomic physicists. These triumphs—of the biochemist and the physicist—are the cumulative products of different disciplines, indeed of different types of human minds. Structural chemistry is based on a system of inference rooted in images; the discipline which opened the core of the atom has a mathematical foundation. Since the skill of the physicist struck terror in our hearts, those who know about it outnumber those who understand it, by legion, but the equally spectacular and potentially more beneficial achievement of the biochemist is known only to the handful who understand it.

8. The Highway to Our Cells: Blood

"Blood is a truly remarkable juice," said Mephistopheles to Faust as the two went through the sealing of their contract.

Mephistopheles showed rare biochemical insight: blood *is* a truly remarkable juice. It is a juice to which we owe much. We owe to it our size, we owe to it our brain, we owe to it our wonderfully complex physiological existence.

It was the great French physiologist, Claude Bernard, who pointed out, in 1878, that the evolution of the highest forms of life has been made possible by the liquid *milieu intérieur*. "The living organism," he wrote, "does not really exist in the *milieu extérieur* [the atmosphere for terrestrial animals; salt, or fresh water for those who had not invented lungs] but in the liquid *milieu intérieur* formed by the circulating organic liquid which surrounds and bathes all the tissue elements."

Complex life is possible for the biological organism only with adequate means of transportation from organ to organ, just as complex social life is made possible only by transportation facilities from community to community.

Blood is the highway to the cells of our tissues. Without it the cells, even those on the surface of our skin, would perish. They would lack oxygen; they would lack food; they would be killed by poisons of their own making.

The center of this most wonderful system of transportation is, of course, the heart. It pumps, in a lifetime of seventy years, about two billion times, and it pushes on its path a hundred million gallons of blood. From the right heart to the lungs, from the lungs to the left heart, from there, through the arteries into the capillaries of the tissues, back through the veins into the right heart, 'round and 'round goes the blood in its wondrous, uninterrupted circle, performing many chores on its rounds. It takes carbon dioxide from the cells and exchanges it for oxygen in the lungs. It is a traveling department store of foodstuffs. It carries everything a cell needs: amino acids, fats, sugars, vitamins, and salts. A hungry cell in one of the outlying districts, say in the toe, extracts from the blood swishing by whatever it requires: a few million molecules of glucose, a hundred thousand molecules of vitamin B_6, a few thousand cobalt atoms. Each cell, however far removed, is thus as well provisioned as a cell in the heart itself.

In addition to foods, the blood carries a variety of other wares: hormones to stimulate laggard cellular mechanisms; antibodies to battle invading poisons; clotting agents to seal breaches in its cyclic path. Furthermore, the blood distributes the heat evolved from the furnace of the cell, thus maintaining a uniform temperature throughout the body.

Finally, the acidity of tissues is kept within tolerable bounds by the blood. Life is fenced in within very narrow limits of acidity. A variety of acids are produced by the metabolic activities within the body cells. Carbon dioxide, lactic acid, and uric acid all tend to acidify the human body. Excess acidity slows down many enzymes, and unless the acidity is counteracted, life itself slows down and eventually halts. Many bacteria—for example, those which turn milk sour—are destroyed by their own metabolism. They produce lactic acid and throw it out into the surrounding fluid. Soon, so much acid accumulates that they become the victims of their own sewage. The blood contains powerful neutralizing mechanisms ready to pounce on the acids cast off by the cells.

How does this "remarkable juice" perform its many functions? It consists of both cells and a variety of noncellular, dissolved materials. About 45 percent of the blood is composed of the red cells. They are tiny red discs (five million of them are packed into a volume the size of a sugar crystal) containing the red protein hemoglobin. This is the truck on which oxygen and carbon dioxide are shuttled back and forth. It is an unusual truck, the hemoglobin molecule, it can carry only one five-hundredth of its weight of oxygen.

A hemoglobin molecule is made of a colorless protein, globin, and an iron-containing pigment, heme. Oxygen unites with this complex molecule in the lungs forming a definite chemical combination which, however, is easily decomposed in the capillaries, liberating oxygen to the gasping tissue cells. Once freed of the oxygen, the hemoglobin forms a temporary chemical alliance with carbon dioxide which is then ferried to the lungs.

Unfortunately, hemoglobin is not very discriminating as it forms its chemical alliances. Life can be snuffed out by carbon monoxide poisoning because of this lack of discrimination by the hemoglobin. It combines with carbon monoxide gas with far greater avidity (two hundred times greater) than with oxygen. Thus, if one inhales a mixture of oxygen and carbon monoxide, the two gasses compete for the favors of the hemoglobin, and oxygen has odds of two hundred to one against it in the contest.

Death from carbon monoxide asphyxiation is caused simply by the lack of free hemoglobin to transport oxygen to the starving tissue cells. The hemoglobin–carbon monoxide combination is actually not poisonous. City dwellers invariably have about one percent of their hemoglobin tied down to carbon monoxide, and tobacco smoking immobilizes as much as five percent of the hemoglobin. (Carbon monoxide is produced by the incomplete burning of the tobacco.) Fortunately, the combination between hemoglobin and carbon monoxide is not permanent. The gas is quickly swept out of the system, unless, of course, enough was absorbed to overwhelm the victim.

Red cells are highly specialized for their role as trucks for the transport of oxygen and carbon dioxide. Their metabolism is very low; they are stripped down to such an extent that the mature cells lack even a cell nucleus. They are nourished by the metabolism of other cells as long as they are useful, but when in old age they falter at their tasks they are liquidated. There are special cells in the blood vessels of the spleen and liver which unceremoniously devour the aging red cells. These cannibalistic phagocytes (eating cells) are constantly trapping and dismembering the more sluggishly moving red cells. The iron is salvaged from the wreckage but the pigment, heme, is piped into the gall bladder, from where it is discharged into the intestine. The protein, globin, is cleaved by enzymes and the amino acids are recycled for the synthesis of other proteins.

Why the red cells must die is but one of the unsolved mysteries connected with them. There are many others. How do the phagocytes select their aging prey for the slaughter? Are there stigmata of age, or do the phagocytes merely fall upon the laggard cells?

All cells, young and old, travel under the impetus of the pumping heart. What makes an aging cell slow down in the capillaries? Is there a definite retirement age for red cells or are they destroyed at random?

Only to the last question do we have an unequivocal answer. The meteoric existence of the red cells was known for a long time, but estimates of their career varied from five days to two hundred days. The life span of the red cell was clocked with isotopes. The counting of the days of the red cell was a serendipitous byproduct of an entirely different project—as so often happens in scientific research.

A biochemist who completes merely his original projects is rather limited. The chemical ways of the cell are so much more complex than we can at present imagine, that, during the course of almost any project, mechanisms of far greater interest than the one originally visualized are invariably exposed to observant eyes. This is why freedom for the investigator to follow up unexpected, chance findings is so essential.

The metabolism of the simplest of amino acids, glycine, was to be studied in humans. Dr. David Shemin at Columbia University made two ounces of glycine which contained not the ordinary isotope of nitrogen but its rare, heavy isotope, and, showing the ultimate in confidence in the purity of his preparation, he promptly ate it. For weeks thereafter he obtained samples of his own blood and measured the heavy nitrogen isotope in the various fractions of it. The largest concentration of the isotope was in the red pigment of the hemoglobin. Apparently this pigment, heme, is fashioned out of the glycine molecule, hence the high concentration of the heavy isotope in it. Glycine, which contains but one nitrogen and two carbon atoms, can be marshaled by the enzymes of the bone marrow, the birthplace of the red cells, into the elaborate structure of heme, which contains thirty-four carbon and four nitrogen atoms. The carbon atoms do not all originate from glycine. A four-carbon component of the Krebs cycle first fuses with glycine and then this newly formed thread is woven into the tapestry of the hemin molecule. This is one more example of the great versatility of the cell in fashioning its components of intricate structure and wondrous variety from abundant small molecules.

The entry of glycine into the hemin molecule pointed a way for the measuring of the life span of the red cell. Only those red cells which are made on the day or two following a meal of labeled glycine contain the heavy isotope of nitrogen. As these cells are devoured at the end of their careers, the isotope-containing heme from them is voided through the bile and lost from the body forever. Therefore the disappearance of the heavy isotope from the heme marks the death and disposal of the cells produced on the day of the isotopic meal.

Small samples of blood were tapped almost daily after a meal of labeled glycine. The amount of heavy nitrogen remained at a constant level for about eighty days. After that it began to disappear. From a mathematical analysis of the complete data we can calculate that the average life span of the human red cell is one hundred and twenty days. The "average" is emphasized, for there was a small amount of heavy nitrogen isotope left even after one hundred and thirty days. Apparently red cells, just like the whole organisms of which they are a peripatetic part, vary in their life span. Some red cells exceed, and some fall short of the six-score days.

The rate of destruction and of synthesis of red cells is prodigious: about ten million red cells are born and about the same number die every second in each of us. The number of red cells can increase for a variety of reasons. One of the most interesting of these is prolonged stay at high altitudes. Since the concentration of oxygen at high altitudes is low, the body adapts itself to the emergency by making more cells and hemoglobin for the transport of the gas in short supply.

The ability to adapt to lowered oxygen concentrations is a valuable asset. For the greater the adaptability to changing environment, the greater are the chances of survival of an organism. The red blood corpuscles are not the only cells which can be augmented to compensate for an environmental or structural deficiency. If one kidney is removed from an animal the other kidney will increase in size, thus enlarging its functional surface and enabling it to carry the larger burden which falls upon it. The mechanism of the induction of the formation of new cells by a chemical stress—a decrease in the available oxygen or an increase in the amount of waste products awaiting disposal—is a challenging problem for the biochemist. For here he is coming to grips with the basic unsolved problem: the method of synthesis of a new cell, complete with its structural components, its enzymes, and its urgency to live. We are just beginning to explore the mechanism of adaptation with the tools of biochemistry.

If the red cells are abnormally low in number or are deficient in hemoglobin, one is said to be anemic. Excessive hemorrhage is the simplest and most easily remedied cause of anemia. Fully one fourth of an animal's blood can be lost with impunity: the loss is made up in two to three weeks. This is the reason for the ease with which we can donate a pint of our blood, which is but one tenth of our total wealth of it. Another cause of anemia, particularly among infants, is lack of iron in the diet. Still other causes are poisons which destroy the bone marrow. Finally, the anemia which at one time was as dreaded as cancer is today, is pernicious anemia. A little over forty years ago this disease was as relentlessly fatal as an advanced case of inoperable cancer is today. But the disease is now mastered. Its conquest is a monument to the joint efforts of medicine and chemistry.

Until 1926 we had no idea of the cause of the disease. Infection, poisons, and cancer were all accused as the possible culprits. The methods of treatment were as varied as they were futile. There was an odd brake on progress against the disease: no experimental animals could be induced to come down with it. We can make dogs diabetic by performing a simple operation; we can transplant tumors; any of the infectious diseases can be implanted into almost any animal; but to pernicious anemia all but the human animal seemed to be immune.

Dr. George H. Whipple and his associates at the University of Rochester decided to study experimentally produced anemia even though it did not resemble pernicious anemia. Dogs were bled copiously and frequently to induce this simple anemia. The goal of the study was to see if we could intercede by some means and acclerate the rate of regeneration of red cells. Various dietary aids were tried and it was found that feeding beef liver to anemic dogs helped their recovery.

After this discovery had been made, the next obvious step was taken by two physicians at Harvard, George R. Minot and William P. Murphy. They fed to their patients suffering from pernicious anemia huge amounts of liver and noted marked improvement within ten days.

As long as the patients were kept on a diet of about a pound of liver a day their improvement continued. Apparently there was something in liver which could protect a patient against the ravages of pernicious anemia.

That is as far as the physicians could go in the search for the cure. At this stage the biochemists took over and began to track down with their specialized searching tools the active principle in the liver. The approach of the chemist to such a problem is, by now, familiar to the reader. Using a variety of chemical manipulations—extraction, precipitation, evaporation—the chemist weeds out the inactive contaminants, testing at each stage the efficiency of his gardening by an assay of the potency of his preparations. Needless to say, the product is expected to become more and more active: a smaller and smaller weight of it should contain most of the original biological activity. Crystallized enzymes, vitamins, hormones, and essential amino acids are the testimony to the effectiveness of these methods of the chemist.

This search, however, was hindered by a particularly difficult obstacle. The only test for the presence of the active principle in the liver was the improvement it brought to human patients suffering from pernicious anemia and, with the easily administered liver therapy, the untreated disease was becoming more and more rare. One leading research center offered free hospitalization and medical care to any patient suffering from the disease in return for permission to standardize the preparations with which the patient was being cured.[1]

Only twenty years after the search began was this particular obstacle removed. During those twenty years chemists managed, despite the lack of convenient testing, to concentrate the active principle several thousandfold. Whereas a pound of liver had to be fed to a patient each day, only one milligram (450,000 milligrams make a pound) of the concentrated material was needed per day when it was given in the form of an injection.

The outstanding biochemist engaged in the purification of the liver factor was the late Dr. Henry D. Dakin, one of the stalwart pioneers of American biochemistry. (The antiseptic solution which he devised, and which bears his name, was the life-saving antiseptic of the First World War.)

---

1. The recognition of the special dietary value of beef liver deprived impecunious students of cheap nutritious meals. Beef liver used to be given away gratis by butchers for cat food; the author recalls delicious dinners of onions and calf liver at five cents a pound for the liver.

A point was reached in the purification of the liver factor beyond which further progress seemed almost impossible. The product, even though enormously concentrated, was still far from pure. Testing the material became well nigh impossible: response by the rare human patients to the various concentrated preparations was almost uniform.

Then, in 1946, a brand new kind of dietary deficiency was reported in the rat. If rats were kept on a diet in which the protein source was alcohol-extracted casein, they failed to grow. (Of course the diet was complete in all the known vitamins.) The missing factor could be found in a variety of foods. It was also present in the commercial liver extract which was being used to cure human pernicious anemia. Let us call this unknown material the "rat-growth factor."

There had been studies on the nutrition of chickens, dating back twenty years, which showed that unless laying hens were fed some protein of animal origin, their eggs did not hatch normally. The factor present in animal proteins and essential for normal hatching was called the "animal-protein factor."

There were then three different unknown dietary factors: the pernicious-anemia factor, the rat-growth factor, and the animal-protein factor. These three apparently different problems were tied together and the three factors were proved to be the same, as soon as a pure crystalline material was isolated. Before that could be accomplished, however, a rapid, consistent, and specific assay for these factors was needed. The assay for one of the three apparently divergent factors was provided by Dr. Mary Shorb. She found a microorganism—*Lactobacillus lactis* Dorner, or LLD for short—which requires for *its* growth the "rat-growth factor."

The use of microorganisms for assay purposes is a development of immense value. We have seen earlier that yeasts and animals need the same vitamins, or that the animo acid methionine is essential in the diet not only of men but of mice and of microorganisms as well. Among the thousands of different microorganisms we can find some which require for their growth any of the dietary essentials of animals. Using them instead of animals for assay purposes is a great advantage.

In order to deplete their reserve of a dietary essential, animals sometimes have to be kept on a deficient diet for months. Indeed a deficiency often does not become apparent until the second generation. It takes months to raise a generation of rats but only minutes to raise a generation of microorganisms, so rapidly do they reproduce. Therefore, assays with microorganisms are a matter of hours.

The principle of such an assay is simple. The growth of bacteria is proportional to the amount of balanced diet available to them. If their diet contains all the essentials but one, that one factor, whether it is a vitamin, or an amino acid or a salt, becomes the limiting factor in the growth of the organism. Growth becomes proportional to the amount of the limiting factor available. For example, if a million bacteria grow on one microgram (one twenty-eight-millionth of an ounce) of vitamin B_x, two million will grow on two micrograms of that vitamin. If now, on an unknown amount of vitamin B_x 1.5 million bacteria will grow, we can then conclude, that the unknown sample contains 1.5 micrograms of the vitamin.

Dr. Shorb's microorganism—LLD—could grow with extracts of the same foods which contained the rat-growth factor. Liver extracts were the best source of the LLD factor. Dr. Shorb suggested that perhaps the LLD factor and the pernicious-anemia factor were the same. If this were so, then here, at long last, would be an assay for the pernicious-anemia factor other than a test on a human patient.

Within a year chemists at Merck and Company isolated the pernicious-anemia factor in pure crystalline form. It was the LLD factor, the rat-growth factor, and the animal-protein factor as well. Since the product was now a definite, pure compound and not just a vague "factor" it was entitled to a new name. Vitamin B_{12} was the name given to the shiny red needles which cure pernicious anemia, enable LLD to grow, permit normal hatching of hens' eggs, and allow rats to grow normally.

Vitamin B_{12} is one of the most potent of the known biologically active materials. It is effective in even smaller amounts than is biotin. As little as 3 micrograms, or one ten-millionth of an ounce, when injected into a human, starts an immediate improvement in the patient's blood picture.

The structure of vitamin B_{12} has been decoded. A remarkable constituent of it is the metal cobalt. That traces of cobalt are essential in the diet of the mammal was previously known. There are a number of such trace elements essential for health. Indeed, the list of elements essential in our nutrition begins to read almost like the chemist's Periodic Table of the elements. Now the specific role of one of them, cobalt, became clear. It is part of vitamin B_{12}.

While knowledge of the functions of the inorganic salts is accumulating but slowly, it is known quite well what elements are present in blood and in what quantities. (It is easier, by far, to assay for an element than to pin down its biological function.) Blood contains the same salts as sea water, and these salts are present in approximately the same ratio. However, blood is fivefold more dilute. In other words, the concentration of each element present in blood is one fifth that in sea water.

That their blood is but diluted sea water—with respect to inorganic salts—is telling circumstantial evidence for the marine origin of animals. The odds are enormously against the chance repetition of the same ratio of the same elements in the sea and in blood. The dilution of the blood of animals has been explained through studies made not by biologists but by oceanographers. Apparently, it is not blood which has become diluted but rather it is the ocean which has become more concentrated since animals arose from that vast aqueous cradle. It is known that the inorganic salt content of the ocean keeps increasing steadily. There have been some interesting extrapolations from the rate of that increase to estimate the era when the ocean had one fifth of its present concentration. It is said that the calculated time coincides with the time animals are estimated, from other lines of evidence, to have evolved from their marine precursors.

In addition to the inorganic salts, blood contains dissolved gases. Of these nitrogen is the most abundant and the least useful. (Oxygen and carbon dioxide are not merely dissolved; they are held in chemical bondage.) Nitrogen seems to have no function other than that of a diluent for oxygen. In the gaseous elementary form nitrogen is completely unusable and can be replaced by another gas, for example helium, in the gaseous mixture that an animal breathes. Indeed, that is precisely what is done with the gases pumped down to deep-sea divers in order to minimize the possibility of their experiencing caisson disease, or the bends. This is a neat application of some simple laws of physics to eliminate a dangerous occupational hazard. Caisson disease is caused by the greater solubility of nitrogen in blood at higher pressures. If the pressure on a diver is quickly released—by too rapid surfacing—the nitrogen dissolved during his stay in the depths is suddenly released, forming bubbles in his blood vessels. At that moment the diver's blood simulates in appearance the contents of a bottle of carbonated cherry beverage from which the sealing cap is suddenly removed. The release of gas bubbles in the veins and arteries causes violent pain, convulsions, and, in severe cases, death. Helium is quite as inert as nitrogen in our bodies but is far less soluble. Thus when inhaled even at great pressures it does not endanger the diver, for too little of the gas dissolves to liberate bubbles at atmospheric pressure.

The freedom of the whale from the bends—this occupational hazard of other deep-sea divers—has puzzled biologists for several generations. According to whalers, a wounded whale can take a half a mile of line to the depths and can then surface with startling speed. An equally rapid ascent from two hundred to three hundred feet would surely release enough bubbles of nitrogen to kill a man. How does the whale tolerate the tremendously greater changes in pressure?

It has been claimed that the whale harbors in its blood stream myriads of a certain species of microorganisms which sop up the inhaled nitrogen of their leviathan host. These microorganisms, like those in the nodules of the clover, were supposed to be nitrogen fixing; such organisms have the ability to convert gaseous, molecular nitrogen into soluble compounds of it. The interpretation offered by the discoverer of nitrogen-fixing bacteria in the whale was that as fast as the whale absorbs nitrogen from its lungs the gas is sponged up by the enzymes of the tiny inhabitants in its blood stream. This would be a fascinating teleological symbiosis, and it ought to be true—but it isn't. The observation could not be repeated. The organisms found by the original investigator must have been post-mortem contaminants. This episode reaffirms that scientists are fallible; Moby Dick is not. The whale keeps its mysteries well hidden.

The noncellular part of the blood, the plasma, contains 7 percent dissolved proteins. This, in addition to the 14 percent of hemoglobin, brings the total protein content of blood to 21 percent. That the hemoglobin is not freely dissolved but is packaged in the red cells is one more example of the remarkable chemical foresight of nature. A 21 percent protein solution would have tremendous viscosity and flow would be impossible. Blood would indeed be "thicker than water." As it is, the hemoglobin is safely wrapped up in the red cells and does not impede the easy flow of blood.

The free plasma proteins contain the clotting mechanisms and the antibodies. Great strides have been made, particularly since the start of the Second World War, in the separation of plasma proteins into their various fractions. The products have widespread clinical uses. Absorbable surgical threads and fibrous packing which stop hemorrhage during surgery are everyday tools of the surgeon, made from human plasma proteins.

These ancillary aids to medicine are but some of the sweet fruits which grew from the evil garden of war. Another unexpected fruit is an understanding and, one hopes, the possible control of one of the diseases of blood, sickle-cell anemia. Sickle-cell anemia is a hereditary disease which expresses itself by the presence of crumpled, sickle-shaped red cells in the veins of the subject. The geographical distribution of the disease is curious; it is very high, 15 percent or over, among the populations of central Africa and southern India.

It is somewhat less frequent among peoples living around the Mediterranean basin. In our own Negro population about 8 percent have some sickle cells in their veins. Fortunately only about 0.2 percent have the abnormal cells in sufficiently large concentrations to incapacitate the patients. The puzzling feature of the disease until recently was that the corpuscles have their normal disc shape while they are in the arteries, but crumple into shapeless bags in the veins. The disease, therefore, is not caused merely by a malformation of the cell membranes but, rather, must be due to an abnormal response either by the membrane or its contents to changes in the concentration of oxygen or of carbon dioxide as the corpuscle exchanges these gases in the lungs.

Dr. Linus Pauling, a physical chemist, became interested in biological problems several years ago. He heard of sickle-cell anemia for the first time while he was serving on a committee appointed by President Roosevelt to study means of advancing medicine. Dr. Pauling very naturally approached the disease as a physicochemical problem. He and his associates isolated hemoglobin from the blood of patients with sickle-cell anemia. They then extracted from the hemoglobin the red pigment, heme, in a pure form and compared it with heme obtained from normal humans. The hemes from the two sources were exactly the same. The protein fraction, the globin, was next purified and was made the target of the battery of critical tests and measurements that reveal differences in proteins. One such test is the measurement of the rate of migration of a protein toward one of the poles in an electrical field. Proteins have the ability to assume either a positive or a negative surface charge depending upon the acidity of the liquid in which they are suspended. (Their ambivalence has earned them the name "zwitter-ions," meaning hermaphroditic, charged particles.) The unbalanced, excess electrical charges on the surface of the protein molecule will determine the direction of migration in an electrical field.

A level of acidity could be reached at which the globin from normal red cells moved toward the positive pole and the globin from sickling cells moved toward the negative pole. In other words, the protein from the diseased cells seems to carry a larger positive charge than the normal protein does. Such an abnormal charge can have profound effects on the protein's ability to bind water or the gases carbon dioxide and oxygen to it, and, also, on the ability of the protein molecules themselves to be bound together. Pauling's discovery revealed the cause of the sickling phenomenon: when the abnormal protein is laden with carbon dioxide in the veins it shrivels in volume and fails to fill out the cell membrane. We even know the molecular site of the protein's abnormality. Dr. Vernon Ingram of Cambridge University addressed himself to this task. He confidently expected that the abnormal protein might have some aberration in its amino acid sequence. A complete sequence study as performed by Sanger loomed as a forbidding task: there are close to 600 amino acid units in a complete hemoglobin molecule. Ingram gambled on a modification of Sanger's approach. He fragmented the hemoglobin molecule with an enzyme—trypsin—which is known to cleave proteins only at restricted sites. Trypsin is a specialist in snipping bonds formed only by the amino acids lysine and arginine. It therefore can produce only chunks of the protein whose size would be determined by the number of amino acids between these two amino acids. The mixture of fragments obtained from normal and sickling hemoglobins was permitted to migrate in an electric field in one direction and was then chromatographed at right angles to their original path. The papers were sprayed with a reagent which yields purple areas where it comes in contact with amino acids or protein fragments. There were nearly thirty spots on each paper and their geographical distribution was almost identical. But there was one difference in the "fingerprints" of the protein fragments from the two sources. One fragment—arbitrarily designated as number 4—from the abnormal hemoglobin had slightly different mobility on the paper than its counterpart from the normal hemoglobin. Fragment 4 was found to contain a sequence of eight amino acids which were identical in the two samples except for one amino acid. In the abnormal hemoglobin a glutamic acid was replaced by a valine.

It is this single substitution that makes the difference between healthy and crumpled red cells. Glutamic acid has an extra acidic group in its structure which is negatively charged at the levels of acidity that prevail in a red cell. This extra negative charge is lacking from the hemoglobin of a patient with sickle-cell anemia. A dearth of but one negative charge can doom a subject to the life of an invalid, and in those cases where both parents contributed a sickling gene—i.e., if the patient is homozygous—early death may be his fate.

That such a fumbling can occur in the weaving of the tapestry of the hemoglobin molecule is a source of terror; that it occurs so infrequently is a source of awe. The precision required for the perpetuation of life is almost beyond belief. A human being has thousands of different species of protein molecules. Each one of these has a unique structure; each has an amino acid sequence all its own. Billions of each of these protein molecules must be made every second of our life to shape and to sustain it. The pattern of each protein must be unerringly duplicated. If there is but one fumble and just one amino acid is misplaced, as in sickle-cell anemia, the red cells crumple shapelessly, causing disease or death. This one blunder, with its patently gruesome consequence, testifies to the unfailing success in the shaping of thousands of other kinds of protein molecules in our bodies. Disease or health, indeed, life or death, hangs on a thin thread no stronger than the link between two appropriate amino acids.

It is awesome to contemplate the extent of damage wrought in an individual by the substitution of but one amino acid. The clinical symptoms accumulate from two different stresses. The body tries to eliminate the sickle red cells by destroying them. This imposes an enormous burden of new synthesis. There is an overactivity of the bone marrow, the heart dilates, and since so much biological construction is a total waste, physical development of the individual is poor and weakness and lassitude are his lot. At the same time, the sickling cells tend to clump and thus interfere with circulation, causing a medical-textbook-full of syndromes ranging from brain to kidney damage.

How could a gene which dooms its possessor to inferior physical

structure and early death survive for thousands of years? For, assuming that it was a single mutation, from the currently wide distribution of the deleterious gene the mutation must have occurred
long ago. Why was it not erased from the pool of human genes by
the elimination of its hapless possessors? At first glance the sickle-
cell syndrome defies Darwin's generalization on the survival of the
fittest. However, this genetic paradox was resolved recently. In
some areas of the earth the sickling syndrome actually confers an
advantage on the person it afflicts! Sickle-cell anemia has the highest frequency in Africa and India, where falciparum malaria is endemic. It was discovered that the protozoan *Plasmodium falciparum*, which is the etiologic agent of this disease, declines to
infest sickling red cells. Sickle-cell anemia thus confers immunity
against malaria. This is a rare case of a deleterious mutation becoming an advantage in a hostile environment.

Interceding with chemical agents to overcome the crippling effects of sickling became a challenge. Someone reported that the ad-
ministration of large doses of urea is of some value in rescuing patients with sickle-cell anemia from extreme debilitation. The
continued improvement of the patients was puzzling: urea is rapidly
eliminated by the kidneys. The therapeutic agent turned out to be
not urea but ammonium cyanate which is produced in small
amounts in water solution of urea. (Yes, it is the reverse of the reaction discovered by Wohler.) Cyanate binds to the sickle globin
and retards the clumping of the protein molecules when they become laden with carbon dioxide.

9. Cell Defense

Many are the foes which attack the cell. The assaulting hordes come in strange shapes and varied sizes and aim a diverse battery of weapons at their target. The cell fights back with sustained valor. If its armed vigilance falters, disaster befalls the cell fortress. The struggle starts at birth and continues relentlessly. There is no quarter, no armistice; only survival or death. That cells do survive is a miracle wrought by their defensive weapons—weapons of great variety and of astonishingly ingenious design.

What particular cell-guard weapons from the well-stocked arsenal are mobilized at any one time depends upon the nature of the invader. The cell can burn up the invading enemy; it can, by means of enzymes, alter the marauder to remove its sting; it can fashion special shock troops which, using their strands of protein molecules, grapple with the invader until it is immobilized.

Let us observe some of the embattled units in this constant warfare.

The best defense against any foe is to prevent its penetration. The body's first line of defense is a tough skin, which, though quite effective, has unfortunately, some weak spots. Bacteria lodge in the pores of the sweat glands and in the hair pits, causing pimples and boils. Poisons and bacteria can pour through the large openings, the mouth, nose, and eyes. The stomach, however is booby trapped against the bacteria: the high concentration of acid in its juices kills most of them.

If a simple, chemical poison enters the body it is handled with effective vigor. The defensive campaign follows a well-defined strategy. The largest route of entry for such poisons is, of course, the mouth. The first maxim of the strategy is: Absorb as little of it as possible. A poison can do us no harm while it is in the alimentary canal. The transient contents of that long tube are actually not part of the body; they can do us good or ill only when they gain admission into the blood or other tissues.

There is a remarkable screening performed during absorption. Before a substance can gain admission into the tissues it must pass the discriminating scrutiny of the cells lining the alimentary canal, and these are remarkably adept at excluding undesirables. For example, humans absorb cholesterol with ease. However, there is an astonishingly similar substance made by plants, sitosterol, which is not absorbed at all.

If a poison passes the selective barrier of the alimentary canal, the appropriate order of the body's high command is: Excrete it! Use the kidney, use the lungs or sweat glands, but excrete it. This is, of course, a selectivity too, but in reverse. Now the foreign substances are preferentially expelled.

If excretion fails, the command is: Burn it up! To study the fate of toxic substances in animals we inject them. In this way we by-pass the forbidding scrutiny of the alimentary canal. If we insert into the muscles of an animal some benzene, most of it is promptly oxidized (mostly to carbon dioxide and water), by the enzymes of the victim, and the animal is rid of the poison.

The enzyme systems which burn up toxic substances are not teleologically designed for this single, self-defensive purpose. These enzymes are always present and are usually performing their normal metabolic oxidations. But if a toxic, foreign substance comes along and happens to fit into the working pattern of an enzyme system, the animal benefits from the versatility of its enzymes.

It would be difficult to visualize how an organism could be equipped to handle any poison it might encounter with enzymes tailored a priori to that purpose. There was a period in the development of biochemistry when hundreds of different organic compounds were administered to animals in order to study their fate in the body. Many of these substances had undoubtedly never existed in the universe until zealous organic chemists strung them together. But still, even though the animal had never, in its whole evolutionary history, encountered these substances, it promptly oxidized them or made extensive alterations in their structure.

During the Second World War there was information that the Germans were manufacturing on a large scale a substance called diisopropylfluoro-phosphate. (Even chemists call it DFP.) Since DFP paralyzes its victims, it was feared that the Nazis might use it as a "nerve gas." DFP was therefore extensively studied by the medical unit of the Chemical Warfare Service and a biochemist of that service found an enzyme in the liver of the rabbit which tears DFP apart. Now, DFP does not exist in nature and undoubtedly it never has. If it were not for Hitler, no rabbit might every have made the unpleasant acquaintance of DFP. Yet rabbits have the enzymes to dismember this rare poison.

There is other evidence that these so-called detoxication processes are performed at random. A poison sometimes becomes *more* toxic after the enzymes finish their alterations of it. The fate of ethylene glycol, the poisonous solvent for sulfanilamide, is a good example of this.

Ethylene glycol (the reader may know it is as an antifreeze), is an alcohol. Many alcohols are poisonous—even ethyl alcohol, which is the least poisonous of them, can be quite dangerous. Let us digress a bit from the fate of ethylene glycol and follow the course of ethyl alcohol in the body.

Ethyl alcohol can reach fatal concentrations in the body from the rapid intake of about one tenth of an ounce of 200-proof alcohol per pound of body weight. In other words, if a man weighing 150 pounds drinks rapidly 15 ounces of 200-proof alcohol, his chances of recovery from his alcoholic stupor are mighty slim. (A variety of factors such as the state of the subject's health and his previous experience in the consumption of alcohol make the exact outcome of the experiment unpredictable.)

Alcohols produce their initial toxic effects by inhibiting the rate of respiration within the cells. The assortment of symptoms which mark a man as being intoxicated can be induced in the most righteous teetotaler, without the use of a drop of alcohol, simply by reducing the concentration of oxygen in the atmosphere he breathes. The scarcity of oxygen will limit cellular respiration and the external symptoms of reduced respiration are the same whether it is brought about by an inhibitor or by diminished supplies of oxygen.

The influence of low oxygen concentration has been studied extensively. The pioneer in these studies, the English physiologist Joseph Barcroft, reported that on journeys to high altitudes in unpressurized planes he has witnessed emotional reactions similar to those experienced after an overdose of alcohol: depression, apathy and drowsiness or excitement and joyfulness, and general loss of self-control. "A person may sing or burst into tears for no apparent reason or be extremely quarrelsome, indolent, and reckless." During the Second World War, some members of the crews of high-flying bombers would take off their oxygen masks on the return trip, for a quick, hangover-less bender.

The fate of all alcohols is the same in the body. They are gradually oxidized. In the case of ethyl alcohol the intermediate stage during the course of the oxidation is the formation of acetic acid, a normal constituent of the body. Acetic acid can be either oxidized further to carbon dioxide and water or utilized as a brick for the assembly of a number of more complex body components. Alcohol, therefore, is really a food. A very limited food, to be sure, since, like sugar, it lacks proteins, vitamins, and minerals.

Now, just as ethyl alcohol is first oxidized to an acid, so ethylene glycol is also oxidized to an acid. But this acid, oxalic, happens to be a merciless poison. The enzymes which accomplish this oxidation doom the animal to quick death. In this case it would be much better for those enzymes to lie low and do nothing to the ethylene glycol. The animal might be able to excrete it slowly through the lungs and kidneys and thus, if the dose is not too large, survive. Oxidation of even much smaller doses brings certain death.

The struggle against our bacterial enemies seems to be more purposeful than the disposal of simple toxic substances. However, even this apparently planned campaign may be really more haphazard than it appears. Bacterial poisons are either proteins or other complex molecules. We are unable, with our present knowledge, to view the intimate mechanisms that accomplish the disposal of these poisons. We only see the ultimate effect and that seems mighty purposeful.

Those parts of the body which are not normally in contact with matter from the outside world—the blood and various other tissues—are free from bacteria. Not statically, the way the inside of a can of evaporated milk is bacteria-free, but dynamically free. If any bacteria penetrate into the inner tissues they are attacked by special shock troops for aggressive defense—the white cells of the blood. They are smaller in number than the red cells and, as their name implies, they are without the red hemoglobin. They have, instead, other specialized equipment—enzymes. The white cells can project tiny strands of their tissues and encircle the bacteria. Once trapped the bacteria are helpless, for the potent enzymes oozing out of the white cells tear them to shreds.

A boil on the neck is a typical battleground in such a struggle against marauding enemies. Some microorganisms lodge near the root of a hair and, finding a warm, cozy nook and a source of food, begin to multiply. As a first reaction to this breach in the body's static defense, the area becomes inflamed. The capillaries become dilated, causing the flow of an augmented supply of blood. Fluids from the blood escape into the area and form a clot, converting the whole into a jellylike mass. A barrier of fibrous tissue is formed around this mass, isolating the infection. Meanwhile the white cells have been gathering on the battleground, crawling through the capillary walls, they pounce upon the infecting organisms. If all goes well, the invaders are killed off in this melee. If not, the infection spreads and the battle is repeated at every new focus of infection. The white cells carry out Mr. Churchill's magnificently expressed strategy: wherever the germs may travel in the blood stream, they are pounced upon and engaged in mortal combat. The white cells get powerful assistance, once the infection becomes very widely spread, from the large phagocytes in the capillaries of the liver, spleen, and bone marrow (the same cells which dispose of the aging red corpuscles). If the agencies already committed to the defense are inadequate and the bacteria are winning in those scattered engagements, the body has still another line of defense—but not against every type of invading organism.

It has been part of man's knowledge for a long time that there are diseases from which a survivor rarely suffers again. Apparently that first attack leaves him with a receipt that he has paid his tax of pain to life and he is spared further visits from the same tax collector.

Induction of a mild case of smallpox in order to gain immunity to a severe attack was practiced a score of centuries ago by medicine men in India. They obtained pus from a patient, with a mild case of the disease and smeared it into a scratch on a healthy person. The practice of such preventive inoculations against smallpox continued in the East but was introduced into Europe only in 1718 by Lady Mary Wortley Montagu, the wife of the British Ambassador in Constantinople. She had her son inoculated by a Turkish doctor. The wide prevalence of the disease spurred others to follow the example set by the courageous ambassadress and the practice of inoculation against smallpox became widespread. During the Middle Ages, to be free of pox marks was considered a mark of beauty. Little wonder that people were ready to subject themselves to the inoculations, even though the outcome was not always predictable. Sometimes the induced disease was a severe case of smallpox.

Chance taught us a safer and equally effective mode of protection against this disease.

Cows, too, are susceptible to smallpox. English farmers knew that cowpox was contagious among humans and considered the rather mild disease as an occupational hazard of dairymaids and others who came in close contact with cows. Who made the first observation that a case of the mild cowpox protects the sufferer from the virulent human smallpox is not certain. Some English farmers have been credited with this astute and profoundly useful correlation. But debates on priority, whether of ancient or of current discoveries, are unprofitable. The discovery is of far greater importance than the ego of the discoverer. Certainly it was Dr. Edward Jenner who established the facts with well-documented evidence and demonstrated how to harvest the benefits of the chance discovery. Since cowpox was known technically as *Variolae Vaccinae*, the purposeful inoculation with cowpox pus to immunize against smallpox came to be known as vaccination.

Vaccination became a very widely applied and gratifyingly effective measure of protection against smallpox. But the mechanism which produced the immunity of course remained unknown. In those pre-Pasteur days even the cause of infectious diseases was unknown. Pasteur demonstrated that some diseases and the immunity to them were induced by the same infectious agent. He did this in a celebrated public, scientific demonstration.

Anthrax was decimating the cattle herds of Europe. Pasteur traced the cause of the disease to the tiny rod-shaped bacilli which were teeming in the blood of the diseased farm animals. He also found that these bacilli could be rendered less virulent by keeping them, for a while, at temperatures much higher than normal body temperatures. (Most bacteria have become so accustomed to the warmth of the animals they inhabit that they live best at that temperature even though they themselves are unable to maintain such warmth.) The heat-treated germs would not kill the animals when injected. They merely made the animals ill. However, after the animals recovered from their mild attack of anthrax they were immune to the virulent organisms. They could shake off doses of anthrax bacilli which would surely kill unimmunized animals.

When the conservative physicians scoffed at Pasteur's claims he arranged a public demonstration which had all the trappings of a country fair. Scientists, physicians, dignitaries, and newspapermen all gathered in a field where fifty sheep, a few bottles of bacterial broth, and Pasteur were the center of attention. His assistants injected twenty-five sheep with a heat-weakened bacterial suspension. Twelve days later those same animals received a stronger dose of infection, consisting of bacilli which were exposed to less rigorous heat treatment. The sheep survived this injection, too. Finally all the animals, including the twenty-five "controls" which until then had been untouched, were injected with the same deadly dose of anthrax. Two days later, twenty-two of the controls were dead and the three others were in their final agony. Every one of the sheep protected with the heat-weakened bacilli was grazing contentedly!

Pasteur thus proved not only that infectious diseases are caused by microorganisms but also that those very germs rouse the animal to rally to its own defense and make weapons to repel future invasions. (Of course, today we know that all infectious diseases are not caused by bacteria. Some, including smallpox, are caused by viruses. But not until several years after Pasteur's death were viruses discovered.)

The nature of the weapons of immunity remained obscure for a long time after Pasteur's dramatic success with the attenuated anthrax bacilli.

The allergies which bedevil so many of us are also our saviors from germs. Outbreaks of hives, sneezing, the necessity for abstaining from certain foods, are relatively small prices to pay for having the weapons with which to combat infectious disease. For the same process which visits upon us the discomforts of allergies brings to us the blessings of immunity to disease.

Allergy means simply an "altered reaction." If we inject into a guinea pig some egg white from a hen's egg nothing unusual will happen. But if after some time we inject into the same guinea pig an identical dose of egg white there will be a violent reaction. The animal's breathing will become labored, it will thrash around, and will finally go into shock, from which—depending on the doses injected—it may or may not recover. The guinea pig has an altered reaction or an allergy to the second injection of egg white. What brings about this violently different response to the second injection?

Animals resist the entry of foreign proteins into their tissues. The proteins we eat do not normally enter our tissues intact. These proteins are broken down by the enzymes of the alimentary canal into their component amino acids and only these are permitted to enter our tissues. From the absorbed amino acids we fashion protein molecules in the image of our own proteins. (In patients who suffer from the various allergies there seems to be a minute amount of seepage of the foreign proteins into their tissues.) Therefore if a foreign protein does enter into the tissues of an animal it means that a stranger is within the gates. The stranger may be just a foreign protein molecule or a whole organism with its foreign proteins. The reaction of the animal is the same to either danger. It begins to fashion shock troops, or antibodies, in an attempt to dispose of the invaders.

Antibodies dispatch the invaders to their doom in a variety of ways. They can dissolve the cell walls of bacteria and the tiny monsters just ooze away; or they merely stimulate, by their very presence, the white cells to greater efficiency. Finally, antibodies combine with the intruding protein or germ, the so-called antigen, to form with it an insoluble particle. Once the foreign matter is thus clumped together the scavenging white cells and large phagocytes dissolve them at their leisure.

For example, in the blood of the guinea pig which has been injected with the egg white there appears an antibody which when added to a solution of fresh egg white curdles it. The violent symptoms of the guinea pig on the second injection are caused by the excessive curdling between egg white and antibody within the animal.

All antibodies are amazingly specific. The antibody from the blood of this guinea pig will not curdle the egg white from a duck egg or a goose egg as effectively as it does the hen's egg white. It is easy to see why such remarkable specificity of antibodies is essential. An animal would be in a dire predicament if it produced antibodies which curdled any protein at random. The antibody might curdle the animal's own proteins.

That antibodies are so specific is a great asset. We can take the antibodies formed against diphtheria by a sturdy horse and fortify a human child with those very antibodies. Or we can tell whether a brown stain on a cloth is ox blood or human blood. When the dissolved stain is mixed with the serum of a guinea pig which had previously received injections of human blood, the formation of a precipitate identifies the stain as human blood too.

The structure of an antibody was determined about ten years ago, as a culmination of 120 years of effort. It was observed by the biochemist Henry Bence-Jones in 1847 that patients who have multiple myeloma—one of the many forms of cancer—excrete an unusual protein in their urine. The Bence-Jones protein, as it came to be called, has very unusual solubility properties in warm water. Unlike any other proteins it precipitates out of water at 50°C but redissolves at 100°.

Almost a hundred years later the antibody was identified as a specific protein by the combined efforts of the Swedish biophysicist Arne Tiselius and the American biochemist Michael Heidelberger. Tiselius had developed a method of separating proteins by placing solutions of them into an electrical field and observing how far they migrated either to the negative or positive pole.

The mobility of proteins in such a system depends on their size and the electric charges on them. He could thus separate the proteins present in blood plasma. Tiselius found a protein with minimum mobility which he arbitrarily called gamma globulin. He was able to pinpoint gamma globulin as an antibody when he found with his American collaborator that the level of gamma globulin increased in the plasma of a rabbit which had been exposed to intensive immunization, and in turn was diminished after the addition of the specific antigen.

In the early 1960s two young American scientists surmised that the Bence-Jones protein which pours out in the urine of patients afflicted with multiple myeloma is probably either the immunoglobulin, as it was renamed, or a fragment of it. One of the two young investigators, Dr. Gerald Edelman of the Rockefeller Institute, isolated enough immunoglobulin from sera of patients with multiple myeloma to purify it and to begin to determine the amino acid sequence of it. This was a task similar to the one undertaken by Sanger but greatly magnified in difficulty: immunoglobulin is 25 times larger than insulin.

Edelman's path was eased by the discovery of the English chemist Rodney Porter that the immunoglobulin can be cleaved into three fragments. And, even more important, by the time Edelman started on his jugsaw puzzle, amino acid analysis became an easy task with the aid of sophisticated technology which had been perfected earlier by two of Edelman's colleagues at the Rockefeller Institute, Stanford Moore and William Stein.

Even with these aids Edelman's task was a monumental one, but with imagination and drive, with which he infected his young associates as well, the job was done.

Immunoglobulin has the most complex structure of any protein examined up to the present. It consists of four chains, two "heavy" or long ones and two "light" or short chains, all of them held together by bonds formed by adjacent cysteines (see Figure 9.1).

Analysis of the amino acid sequence revealed an extraordinary difference from all other proteins. As we stated earlier, the absolute fidelity of the amino acid sequence is the hallmark of all proteins; one misplacement, as in sickle-cell hemoglobin, can be disastrous. In the Bence-Jones protein there are 214 amino acids. From amino acid 1 to 108 they are the same in every normal human, from 109 to 214 they are variable. In the heavy chains there is also a variable region of the same length as in the light chain and, of course, a much larger constant chain. The variable chains confer the individuality on the immunoglobulins by means of which they can bind to a vast variety of antigens; the constant chains are needed for other functions, for example to transfer the mother's antibodies across the placenta to the embryo.

The origin of the constant chains is obvious: Since their sequence of amino acids is invariable, and, as we shall see in a later chapter, such invariable sequences are encoded in the genetic material, the ability to form the constant chains must be part of the total biological inheritance of the organism. The origin of the variable chains is baffling. They are apparently formed under the influ-

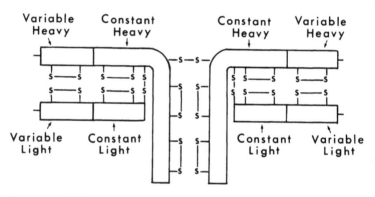

Figure 9.1. Schematic Representation of Immunoglobulin

ence of a challenge by some substance which the immune system recognizes as foreign.

Since these antigens, either natural or man-made, can be almost infinite in number it is improbable that there is preformed information to cope with each one. We must assume that there is the capability to feed back information about the structure of the foreign substance so a sequence of amino acids can be synthesized which can effectively embrace and curdle it.

The mechanism of this feedback system is totally obscure; there are some conjectures, but none of them can be proved or disproved in the laboratory.

The persistence of our antibodies long after the disappearance of the inciting antigen is baffling. We have seen earlier that the body is in a constant state of flux. Tissues are broken down and rebuilt constantly. The amino acids which compose our proteins today will be gone tomorrow and replaced by more recent amino-acid arrivals from our food. Are antibodies an exception to this constant building and dismantling? They are known to remain in the body for years after the infection which caused their formation has subsided. Indeed, sometimes they remain for a whole lifetime. Are antibodies permanent islands in the constantly changing sea of the body?

Dr. Michael Heidelberger, the foremost immunochemist, answered this question in collaboration with the late Dr. Rudolph Schoenheimer. They found that antibodies are no different from other proteins of the body. They, too, are being constantly assembled and dismantled. In other words, the machinery which is set into motion at the time of the original infection to produce antibodies continues its production, sometimes for years, sometimes for a lifetime.

This can only be interpreted by assuming that some cells that accept the specific challenge to produce immunoglobulin become programmed for its synthesis and keep on doing it.

Even though there are large gaps in our knowledge of immunology, the practical achievements in this field are enormous. The multitude of different immunizations, the classification of human blood into various types to insure safe blood transfusions, the diagnosis of diseases (such as syphilis) by testing a few drops of blood, the whole branch of medicine treating the allergies are the bounties harvested from the studies of the antigen-antibody relationship.

Bacterial warfare refers to the use of bacteria or their poisons as weapons in human warfare. But bacteria, too, have been carrying on warfare on a vast scale, millions of years before man and his puny wars became part of the earth's landscape. Man is a fumblng novice in the techniques of mass slaughter. What is it, a hundred thousand humans that we killed with one atomic blow? In less time than it takes to say "hydrogen bomb," millions of bacteria are the casualties in the warfare that goes on in a handful of garden soil. And the weapons used are at least as ingenious as man's.

In that handful of soil are more microorganisms than there are humans on the face of the earth. Life is hard in that handful of a world. Food is scarce; the struggle for survival is fierce. Some of the combatant microorganisms are especially well equipped in their battle against their competitors. They pour out a poisonous solution all around them. In the area staked out by the spreading molecules of the poison, no other organisms can enter and live. The wielder of the poisonous weapon can grow and multiply in his befouled homestead.

It is almost superfluous to say that it was Pasteur who first observed this Lilliputian chemical warfare. A batch of the anthrax bacilli which were described a few pages back stopped growing. He pinned the responsibility for the mass murder on some stray microorganisms which drifted into his cultures from the air.

In the garden of the mind of that genius, this chance observation became the seed of a dream. He visualized the slaughter of the pathogens which cause our diseases by the introduction of their natural enemies or their products into the patient. He wrote in 1877 that such a scheme "justifies the highest hopes for therapeutics."

It took sixty years for that dream to come true. There were little episodes which kept the dream alive during those years. In 1885 Cantani tried to cure a tuberculous woman by the inhalation of a nonpathogenic organism. He was apparently fairly successful. (It may have been a case of a normal recovery from tuberculosis.) At any rate the work was not followed up.

However, many bacteriologists continued to notice the lethal antagonism of some microorganisms to each other. The disappearance of pathogens from water which trickles through the soil became well known.

The destruction of one microbe by another came to be known as antibiosis. At the start of this century the first antibiotic—a substance extracted from one microorganism to kill another—was prepared. However, it was not successful.

The credit for the first isolation of an effective antibiotic goes to Dr. René J. Dubos, a bacteriologist, of the Rockefeller Institute and to his biochemist associate, Dr. R. D. Hotchkiss. Unfortunately their antibiotic, while very effective against bacteria, is also quite poisonous if injected into the patient. It is not used widely, except for surface applications, nor has it received the wide public recognition it deserves.

Dubos's work is the model in the search for all antibiotics. He took a pinch of soil and sprinkled it onto a glass dish coated with nutriment for bacteria. He separated the various strains of bacteria which grew and replanted them into little bacterial gardens of their own with lots of food. Having feasted them, Dubos put the bacteria to work. He placed a growing culture of pathogenic organisms—staphylococci, which cause boils—into each colony of soil bacteria. One strain of the soil bacteria refused to share its food with the staphylococci. These bacteria exude something which kills competing organisms.

The isolation of an antibiotic, later named gramicidin, followed the usual pattern of such searches. The poison-bearing bacteria were grown in large batches and the antibiotic was concentrated from their juice, using the increasing toxicity of the preparations to the staphylococci as the guide in the various steps of the search. The task was completed in 1939. However, the antibiotic turned out to be quite poisonous when injected into animals. Nevertheless, it is a valuable aid in the treatment of exposed infections. But there were other, better antibiotics to come.

Streptomycin was dug out of bacteria by Dr. Dubos's former mentor, Dr. Selman Waksman of Rutgers. This is a rare case of the master following in the footsteps of his disciple.

Waksman has devoted his life to the study of soil microorganisms. They are of enormous importance agriculturally, economically, and even aesthetically. About a ton of leaves falls on each acre of forest every year. If it were not for the soil organisms which decompose all such debris, our earth would become a cluttered, uninhabitable graveyard in a very short time. No dead plants or animals would decompose. The substance of their bodies could never be returned to usefulness. Our accumulation of ancestors for thousands of years back would be with us perfectly preserved.

Waksman had been interested in soil organisms from the point of view of their value in agriculture. But after Dubos had extracted an antibiotic from such organisms, Waksman, too, channeled his efforts in a similar direction. Streptomycin is but one of the many antibiotics he and his associates extracted.

And now a few words about the greatest antibiotic of all those that have been isolated thus far—penicillin. In 1928 a mold spore drifted from the air onto a bacterial culture plate of Dr. Alexander Fleming, a bacteriologist at St. Mary's Hospital in London. The spore reproduced on the spread of food in the glass dish. As it grew, it exuded something, for there was a halo around the mold, an area free of the staphylococci which were the original inhabitants of that plate. Now, such accidents must have happened to scores of bacteriologists. But they would simply throw a ruined plate into an antiseptic bin.

Fleming had been carrying on bacteriological warfare—the proper kind, *against* bacteria—for a quarter of a century. He became interested in this mold which could parachute into a colony of pathogens and could slaughter them with ease. He transferred the mold to culture plates and grew it in large batches. The mold-free juices he prepared were still lethal to bacteria. Since the mold was a *Penicillium* Fleming named the antibacterial substance in the juice penicillin, after the parent. Fleming recognized immediately the potential value of penicillin. He wrote in 1929, that "it may be an efficient antiseptic for application to, or injection into areas infested with penicillin-sensitive microbes." But, as he wrote later, ". . . I failed to concentrate this substance from lack of sufficient chemical assistance. . . ."

Not until ten years later was penicillin concentrated and used in human therapy. Dr. H.W. Florey, a pathologist at Oxford, undertook in 1938 a systematic search for antibiotic substances. His bio-chemist associate was Dr. Ernst Chain, a refugee from Hitler. The team of Florey and Chain was spectacularly successful in winning penicillin. Their methods were communicated to American labora-tories and pharmaceutical houses under the auspices of the Office of Scientific Research and Development. "The Americans," wrote Fleming, "improved methods of production so that on D day there was enough penicillin for every wounded man who needed it. . . ."

There is beautiful, fairy-tale justice in Chain's career. Exiled from his home, he became a key man in the fashioning of a drug which protected millions of young champions as they sallied forth to liberate his homeland.

How does this most potent of cell defenses—which we borrow from molds for our own defense—kill bacteria?

The answer came slowly and piecemeal—as do all answers to questions in biology. First of all it was found that penicillin is lethal only to growing bacteria. If a bacterium is not growing, because of the lack of some nutrient, then it can survive exposure to penicil-lin. Dr. Hotchkiss, whom we met a few pages ago, made the first observation that bacterial cultures in the presence of penicillin ac-cumulate amino acid complexes. Then a young chemist, Dr. J.T. Park, working at the U.S. Army Bacteriological Warfare Research Center, made a penetrating observation. The amino acid com-plexes excreted by bacteria which Hotchkiss observed were attached to a nucleic acid fragment. Meanwhile other biochemists, mostly in England, whose interest was the structure of the cell walls of bacteria, found after painstaking analysis that the cell walls of bac-teria are composed, in part, of the same complex which ac-cumulates in the presence of penicillin. The mosaic now started to fit into a picture: penicillin causes cell wall components of bacteria to accumulate. The next large piece of information was contributed by a Swede, Dr. Claus Weibull. He was studying the mechanism by which an enzyme called lysozyme destroys bacteria. This enzyme which can dissolve bacteria was originally discovered as a component of tears by Dr. Fleming himself. This story has enough improbable coincidences to sound as if it were a plot spun by Dickens. But, actually, this is not unusual in the history of science, since those engaged in scientific pursuits have been so few. In turn, those who have made lasting contributions are fewer still, and, therefore, several achievements by the same person in the same area are not unusual.

Dr. Weibull observed that in the presence of lysozyme bacteria just vanish in an ordinary culture suspension. However, if he supplemented the culture fluid with large amounts of salt or cane sugar the bacteria did not vanish but merely lost their shape; instead of their characteristic rod-shaped structures they appeared as shapeless globules. This finding was a clue on how to observe penicillin at work. The one who recognized the clue is Dr. Joshua Lederberg, one of the more brilliant of contemporary microbiologists.

Dr. Lederberg allowed penicillin to attack bacteria in a culture medium into which he had incorporated cane sugar, after Dr. Weibull's example. He observed that in the presence of the sugar bacteria did not dissolve away as they usually do but merely assumed altered shapes; they too became round globules. If these globules were removed from the environment of penicillin, rod-shaped structures slowly emerged from them. The mode of action of penicillin now became clear. A bacterium is but a minute speck of protoplasm awash in a vast, hostile sea. To guard the precious protoplasm against dissolution and to give them some structural strength bacteria have evolved cell walls. Penicillin interferes somehow with the production of those walls, that is why the cell wall building units accumulate in their culture fluids. Unaided by their walls the newly growing bacteria pop like bubbles in a churning sea. Sugar or salt can protect the bacteria by counteracting, through a physical phenomenon, the pressure accumulating within the naked bacterial globule. It now became clear why only growing bacteria are killed by penicillin. A bacterium whose cell wall is fully formed is beyond the reach of penicillin which can kill only by preventing cell wall formation.

The reason for the complete innocuousness of penicillin for animals also became apparent. Since animal cells have no walls, penicillin can do them no harm. (The rare complications in penicillin therapy are due to allergic reactions.)

Penicillin is truly a wonder drug. But its wonder does not lie merely in its therapeutic effectiveness. Of equal wonder is the finding of a drug which pounces with awesome efficiency on a rare difference in bacterial and mammalian cells, the presence of a cell wall.

We know the chemical structure of penicillin and have even made it in the laboratory. How simple that sounds! That brief sentence summarizes years of work during World War II by scores of the best organic chemists of England and America. This Allied team was led by two of the ablest generals of organic chemistry of the two countries: Sir Robert Robinson, who came to this campaign after brilliant victories in the field of the structure of natural pigments, and H.T. Clarke, who decoded the structure and achieved the synthesis of the sulfur-containing part of vitamin B_1. This cooperative effort is one of the finest examples of programmed research. Florey and Chain developed the methods of isolation and clinical testing of penicillin; Robinson and Clarke decoded its structure; and based on this information DuVigneaud synthesized it. Four of the above who pooled their talents to a common good were Nobel Prize winners. The War was the spur for these unusually gifted people to lay aside their own pet research projects for the duration. But such collaboration on a targeted goal is needed in our continuing struggle against disease. After basic research brings us to a certain level of understanding of a problem, its exploitation to our advantage is often most effectively carried out by targeted group effort. Research support must be judiciously distributed to nurture basic research and to harvest its fruit as expeditiously as possible.

While the synthesis of penicillin is a brilliant achievement of this international team, so far it has no practical significance. It is cheaper by far to let the molds do their chemistry and make it for us.

We do not know the specific function of penicillin within the mold itself. Penicillin may be a normal metabolic product of the mold which happens to be poisonous to other microorganisms; or, it may have no role within the cell other than self-defense. A chance mutation, resulting in the ability to make penicillin, may have enabled that particular mold to survive, for nature achieves her purpose in many apparently purposeless ways.

The parallel history of penicillin and of Dubos's antibiotic illustrates the role of chance in the rewards of research, too. Two groups of scientists, Fleming, Florey, and Chain on the one hand, and Dubos and his team on the other, set out on a hunt for an antibiotic like so many prospectors hunting for gold. The prospectors have the same training, the same skills and tools. The team which succeeded first, later found fool's gold mixed with the gold: the antibiotic was toxic. The other team, by chance, found pure gold, penicillin, and received accolades and Nobel Prizes. But, well has it been said that "The object of research is the advancement, not of the investigator, but of knowledge."

God is known first through nature.
GALILEO, 1615

10. The Identity of the Gene Revealed

"Like begets like." This truism has been part of man's armory of knowledge from a time beyond the reach of estimate. To be sure, spinners of tales might leap over the fence of reality into realms of fantasy where dragons would grow from innocuous seeds and beanstalks would defy all restraints and reach to the sky. But the farmers of antiquity planting their grains in the valley of the Nile, the Euphrates, and the Yangtze confidently expected to harvest not dragons or monstrous beanstalks but multiple replicas of their precious, planted grain.

Even more subtle expressions of biological inheritance were understood by primitive peoples. Arabian racehorse breeders kept elaborate pedigree records of their racehorses for centuries. But while it was recognized that certain traits which can vary among individuals are passed down from parent to offspring, the patterns which heredity follows were unknown until very recently.

It is odd how knowledge of ourselves consistently lags behind knowledge of our outer world. Johannes Kepler perceived the orbital motion of the planets over 350 years ago. We confirmed his hypothesis only a score of years ago when we achieved man-made, unfettered orbital motion. Yet this remarkable genius who, with rare perception, could bring order out of the apparent chaos of celestial bodies could not escape the prison of ignorance in contemporary biological lore: he believed that fish could arise by spontaneous generation from the salt water of the seas just as comets arise in the skies. And in the latter half of the nineteenth century, when we were the masters of the seas with palatial steamships, we could not even guess at the probability of the birth of a blue-eyed baby to brown-eyed parents.

For example, Darwin, whose hypothesis on evolution was based on the assumption of the existence of heritable variations among individuals of the same species, knew nothing about the patterns that heredity follows. In 1872 he wrote: "The laws governing inheritance are for the most part unknown. No one can say why the same peculiarity in different individuals of the same species, or in different species, is sometimes inherited and sometimes not so; why the child often reverts in certain characteristics to its grandfather or grandmother or more remote ancestor."

Ironically, the answers to the questions Darwin was seeking had been known for six years. But the work of the brilliant amateur biologist who observed the pattern—indeed the laws—which heredity follows lay unread and unappreciated in a dusty volume of the *Proceedings of the Natural History Society of Brunn.*

The obscure biologist was Gregor Johann Mendel, who was born in 1822 in Silesia to peasant parents. After finishing his secondary education he endured economic and physical hardships and decided to enter a profession in which "he would be freed of the bitter necessities of life." Thus, at the age of twenty-one, he became a monk.

After his ordination he was assigned to duty as a temporary teacher and remained in this category because he kept failing the examinations for a full-fledged position. It has been suggested that Mendel was stimulated to start his experiments by a dispute with his examiner in botany on the last of his unsuccessful examinations. Whatever were his motivations, he pursued his studies on plant hybridization in the tiny garden of the monastery of St. Thomas at Brunn, Austria, with such devotion and brilliance that he was lifted from his self-imposed anonymity and is counted among those men of rare genius who are the first to discover a law of Nature.

He studied the effects of the crossbreeding of two plants endowed with contrasting traits. He started by planting different peas, tall and short, in the monastery garden. If these two varieties of peas were allowed to self-fertilize, with no possibility of cross-fertilization, the plants yielded seeds which bred true to type: they grew into plants, tall or short, like their parents. But if a tall and a short pea were crossed-fertilized or "hybridized," the seeds from such a union gave rise only to tall peas. Mendel did not stop there. He patiently crossed these tall peas of mixed ancestry and collected their seeds. Out of these seeds grew both tall and short peas. He carefully recorded his seeds and crops and found from over a thousand different plantings that the tall and short "grandchildren" always appeared in a definite ratio: 75 percent tall and 25 percent short. He repeated these studies with red- and white-flowered peas. When he cross-fertilized these flowers he found that all the seeds gave rise to red flowers. But this new generation of red flowers, when crossed-fertilized, produced seeds from which grew both red and white peas. He found 6022 red and 2001 white flowers, again a ratio of three to one (75.1 to 24.9 percent, to be more exact).[1]

Mendel concluded that there are factors in peas which determine their color and height. The factor for whiteness or shortness remains dormant in the first generation after the crossbreeding of opposing traits, but asserts itself in the second. Moreover, the dormant factor reappears in the second generation in a ratio of one to three of the dominant factor.

So here for the first time the pattern of inheritance was revealed. Characteristics of the two parents are not transmitted to the off-

1. The fatuous accusation has been made that Mendel's data are so neat he must have manipulated them. Such an accusation is stupid for two reasons. You can manipulate data only if the expected values are known. Mendel had no known predecessor. He was a unique pioneer. Moreover, in those days there were no rewards for new knowledge. Mankind could not care less about the garden putterings of an obscure Austrian monk.

spring haphazardly but by some directing mechanism which achieved sufficient accuracy to entitle the pattern to be called a law.

Mendel was aware of the importance of his discovery. He tried to get fellow scientists interested in it and sent a copy of his findings to an outstanding Swiss botanist, Karl von Nägeli. But the latter had his own ideas on the mechanism of heredity and brushed aside the presumptuous claims of an obscure amateur.

So the meticulous report on the results of Mendel's eight years of work was buried in the pages of the provincial journal where they were printed in 1866. Two years later Mendel was elected the abbot of the monastery and, as has happened to a good many scientists since then, he abdicated science and became an administrator. He died in 1884 completely neglected by the scientific world which was to discover him only sixteen years later.

The discovery of Mendel's work was made simultaneously by three different investigators who through studies of their own arrived at the same conclusions as did the patient monk.

The three, Hugo de Vries, who was a Dutch botanist, Carl Correns, a German botanist, and Erich von Tshermak, a Viennese plant breeder, apparently learned of Mendel's work from a reference in a comprehensive bibliography on plant hybridization compiled in 1881 by some meticulous German scholar. Each of them graciously acknowledged Mendel's priority on the discovery of what they designated as "Mendel's laws."

What were these laws? In the first place Mendel determined that a single pollen achieves fertilization. (That is, of course, also true of fertilization among animals where only one sperm can penetrate an egg.) Mendel set the pattern for studying the paths of heredity: one must choose a single pair of easily recognizable, contrasting characteristics, e.g., tall, short. One of these may turn out to be a *dominant* and the other a *recessive* trait.

Recessive traits vanish from the appearance of the second generation, only to reappear in the third generation in a ratio of 1 recessive to 3 dominant.

Finally, Mendel assumed the existence of a "formative element" in each pollen and each egg which is capable of determining a single characteristic, e.g., shortness or whiteness, in the offspring.

Mendel's experiments have stood the test of countless repetitions with every species of living organisms that reproduce by the fusion of two sex cells. Every creature from man to mouse shows recessive and dominant traits and the expression of these traits usually follows Mendel's law.

During the hundred years following the publication of Mendel's findings we have slowly pieced together the molecular mechanisms which unerringly achieve the transmission of heritable traits to the offspring.

The pioneer who unwittingly made the first contribution toward the unraveling of the molecular mechanism of inheritance was the Swiss scientist Friedrich Miescher. His work was contemporaneous with Mendel's and without realizing it he isolated for the first time the component of the cell which turned out to be the "formative element"—the hypothetical substance to which Mendel intuitively ascribed the capacity to express a heritable trait.

Miescher was interested in the makeup of the nucleus within the protoplasm. He realized that looking at a cell's nucleus through a microscope, after a variety of stains were applied to the tissues, was not enough, so he traveled to Tubingen, Germany, to study with Professor Felix Hoppe-Seyler, who was an expert in what was then called medicinal chemistry.

It was known from microscopic examinations that the nucleus made up an inordinately large proportion of a pus cell. So Miescher, with Swiss patience, started to collect the oozing surgical bandages which were peeled off purulent wounds in the hospital of Tubingen. He soaked off the organic material, digested the product with the digestive enzyme of the stomach, pepsin, and subjected the mixture he obtained to a variety of chemical manipulations. That he did not have an easy time of it was obvious from his contemporary comment: "I feel as if I am mired in a swamp."

Finally his exertions yielded a material which, unlike other proteins then known, was not soluble in water or dilute acid but did dissolve in dilute alkali. Its difference from other proteins was marked even more when it was found to contain phosphorus. This element had until then been found only in one other constituent of the cell: Hoppe-Seyler himself found it in a component of fats.

Miescher named the novel substance he found in nuclear material "nuclein" and wrote up his efforts in a scientific paper and sent it in 1869 to Hoppe-Seyler to publish. That gentleman edited a journal, which he had named in a burst of modesty *Hoppe-Seyler's Journal of Medical Chemistry*. Hoppe-Seyler was a suspicious editor and would not publish Miescher's paper until he himself had repeated and confirmed the surprising discovery. So two years later Miescher's original account, plus Hoppe-Seyler's confirmation, plus two new papers on the same subject by two of Hoppe-Seyler's students appeared simultaneously.

These uninvited fellow travelers of Miescher's paper showed that nuclein is not restricted to pus cells but is present in the red cells of animals, in yeast cells, and in casein from milk.

Miescher hit upon a more congenial source material for his continuing studies of nuclein. Basel is on the river Rhine and salmon battle their way up this river to their fresh-water spawning grounds. The sperm of salmon is hardly more than a nucleus with an apparatus of locomotion, the tail attached to it. Salmon sperm became the main source for Miescher's studies—a welcome change from the stinking surgical bandages of Tubingen.

While studying the sperm of salmon Miescher made another penetrating observation. Salmon do not eat during their fresh-water pilgrimage to their spawning grounds. During the voyage the musculature of the salmon diminishes while, concomitantly, their sperm content burgeons. Miescher suggested that in these fish, muscle must be converted to sperm. In view of the primitive state of knowledge about body chemistry at that time this was a prophetic observation on the interconvertibility of body constituents. Not until half a century later could we prove conclusively by means of isotopic labels the validity of Miescher's penetrating conclusion. He intuitively sensed the importance of nucleins. He wrote: "A knowledge of the relationships between nuclear materials, proteins, and their immediate products of metabolism will gradually help to raise the curtain which at present so completely veils the inner process of cellular growth."

After Miescher one of his successors in the field coined a new term which competed with "nuclein" so successfully that the latter retains only a historical interest whereas "nucleic acid" today is almost a household word. For by now we know that every living organism from the mighty whale to the tiniest speck of virus particle is built from a blueprint of nucleic acid: The secret of the mechanisms of heredity whose pattern was first glimpsed by Mendel is locked in the wondrous structure of the substance first isolated by Miescher.

A living cell functions on the harmonious cooperation of perhaps 10,000 different enzymes. In turn, an organism such as man is a constellation composed of a hundred thousand billions of such cells.

How are these enzymes made, and what happens to the whole organism if there is a failure in the making of one or more enzymes? The first, the more fundamental of the two questions, was actually answered about seventy-five years ago, but we were too obtuse to recognize it.

In 1902 a keenly observant English physician and biochemist, Sir Archibald E. Garrod, published a paper which he entitled prophetically: "The Incidence of Alkaptonuria, a Study in Chemical Individuality."

The visible diagnostic characteristic of alkaptonuria is the blackening of the patient's urine when stale. (The telltale darkening of diapers or bed linen permits easy detection during infancy.) Some of the other manifestations of the disease are the pigmentation of cartilage and tendon tissue in the adult.

The agent, which turns black on prolonged contact with air, was isolated and proved to be a chemical called homogentisic acid. Homogentisic acid is present in the urine of these patients but it is absent from urine voided by normal humans. Garrod traced the genealogy of some of his patients and found that alkaptonuria is an inherited disease. It is passed down from parent to child as a recessive Mendelian character. Garrod recognized that the symptoms are the result of some aberration of metabolism and called the disease, very aptly, an "inborn error of metabolism." Homogentisic acid proved to be a normal intermediate in the metabolism of the amino acid tyrosine.

For reasons which were unknown at the time Garrod made his discovery, a patient afflicted with alkaptonuria cannot metabolize homogentisic acid to carbon dioxide and water, a feat which a normal human achieves with the greatest of ease. But Garrod made a penetrating conjecture on the source of the trouble in alkaptonuria. He wrote as early as 1923: "We may further conceive that the splitting of the benzene ring of homogentisic acid in normal metabolism is the work of a special enzyme, that in congenital alkaptonuria this enzyme is wanting." In other words, since hereditary diseases stem from the gene, the failure of the gene in alkaptonuria is the failure to fashion the appropriate enzyme. This was the first statement of the seminal relationship between gene and enzyme. However, the classical geneticists were not yet ready to think in terms of molecular mechanisms. They did not visualize the gene as a specific agent but rather as some vague vitalistic entity. Therefore, the possibility that the gene is a molecular component of a cell which shapes an enzyme was not ready for acceptance.

It frequently happens in the history of science that a novel interpretation or observation leads to a host of similar findings because the attitudes of other investigators are reoriented by the perceptive pioneer. Garrod's patient research and intuitive interpretation was such a turning point. A number of different inborn errors of metabolism have been catalogued in the past seventy years since Garrod's initial observation. The ones that are recognized are those in which the crippling effects of the metabolic blunder are relatively minor, enabling the afflicted to survive after birth into infancy or even adulthood. However, there must be many others which are unrecognized because they doom the foetus or the newborn to death of undiagnosable origin.

The extensive, unexpected damage that an inherited metabolic error can produce is well illustrated by a syndrome which is variously called phenylketonuria or phenylpyruvic oligophrenia. These patients, too, excrete in their urine a substance, phenyl-pyruvic acid, which is largely absent from normal urine. If they survive infancy, as many do, the skull of the afflicted does not achieve normal growth—hence the term oligophrenic. The brains housed by the tiny skulls function but poorly, resulting in intelligence quotients of 20 to 50.

Research showed that the metabolic blunder is a block in the patient's ability to convert one amino acid, phenylalanine, to another, tyrosine.

The biological reaction which goes with ease in a normal human goes on at only one-tenth the normal rate in those afflicted with phenylketonuria. The excretion of phenylpyruvic acid results from an alteration effected by the kidney in the accumulated phenylalanine.

Since tyrosine is the source of the body's pigments, these people are usually blond because of the lack of adequate source material. From the clinical point of view phenylketonuria turned out to be a milestone because it is the first mentally crippling disease whose destructive effects can be partially obviated, provided the diagnosis is made early. It was found that if the afflicted infants are placed on a diet that is very low in phenylalanine they escape some of the brain damage.

The inborn errors of metabolism should have served as a clear clue to the mechanism of gene action. They are hereditary diseases, therefore they are transmitted by the presence, or, more likely, the absence of a gene. If a phenylketonuric cannot convert phenylalanine to tyrosine it must mean that the appropriate enzymes are lacking, since all biochemical processes are achieved by those wondrous catalysts of the cell. But this is obvious in hindsight only. And hindsight, contrary to real sight, sharpens with the passing years.

Not every enzyme is of pivotal importance for the life of an organism. For example, we humans have lost the necessary enzymes for the manufacture of nine out of twenty amino acids. During our evolutionary ascent we also lost the enzymatic capabilities for the shaping of vitamins as well. The loss of those genes and enzymes was not deleterious, for having become predators, we could easily compensate for them by eating other organisms, plants, and animals that retained these synthetic capabilities.

However, with advanced cultural development, as we humans become less proficient predators and failed to compensate for it by increasing the sources of food with adequate farming and animal husbandry, those lacunae in our enzymatic armory became increasingly felt. That primitive peoples suffered from nutritional deficiencies is testified to by the stigmata of the diseases in their remnant bones and by some of the stories in their religious lore. But only within the past seventy years have we correlated certain diseases with the deficiency of vitamins in the diet. Still more recently are we aware of the organic symptoms of malnutrition induced by amino acid deficiency.

Obviously all gene losses and, consequently, their enzymes are not equally harmful. The loss of enzymes for the making of vitamins has not been an insurmountable blow. (Especially is this true today when we can manufacture vitamins by the hundreds of tons.) Nor is the loss of the ability to form body pigments incapacitating. Albinism, which is prevalent among all mammals, is but a minor handicap among humans. (They sunburn too easily.) Among whales albinism is probably no handicap at all. (If Melville is to be believed, Moby Dick was huge in size, prodigious in endurance, and fierce in temper. For despite what the literary psychoanalysts try to make him out to be, Moby Dick was just an albino whale.) The loss of the enzyme which converts phenylalanine to tyrosine in phenylketonurics is a more severe handicap. The accumulating products do irreparable damage to the brain. The more peripheral the function of an enzyme, the less damaging is its loss to an organism. The disappearance of a pivotal enzyme dooms the creature in which the mutation occurs. Such mutations are lethal and therefore not perpetuated. Obviously, no organism could survive a moment if it lost the ability to make adenosine triphosphate—the ubiquitous source of energy for all biochemical processes.

The information from inborn errors of metabolism failed to reveal to us the relationship between genes and enzymes, but fortunately biological truth has a way of struggling to the surface provided there are gifted men to help it. One such man was Dr. H.J. Muller.

In order to increase the very low rates of natural mutations—which are on the order of a few per million—Muller's predecessors exposed organisms to alcohol, lead, and other toxic substances. The results were either negative or of doubtful significance. Muller decided on stronger measures. In 1927 he exposed the fruit fly to the intensely penetrating energy of X rays.

He was successful beyond his dreams. He found that the spontaneous mutations which had been observed previously among populations of fruit flies—identified, for example, as "miniature wing" or "forked bristles"—appeared with much greater frequency among the offspring of populations exposed to X-rays. But more than this, he found numerous mutations theretofore unobserved, such as "splotched winged" and "sex combless" flies.

The gene was penetrated at last. The intense energy of the X ray on impact with a gene target dislocates its structure and thus produces permanently monstrous offspring. For Muller found that the visibly altered structures in the progeny of the flies exposed to X rays continued in evidence in subsequent generations. The X rays brought about a permanent genetic alteration, in brief, a mutation.

Important though Muller's discovery was, it still told us nothing about the mechanism of gene action. A concatenation of several factors might yield a fly with "miniature wings." But Muller's discovery eventually enabled us to study at the molecular level how the invisible gene controls the shape, the color, the very essence of a living thing.

A brilliantly successful partnership between Dr. G.W. Beadle, a geneticist, and Dr. E.L. Tatum, a biochemist, was formed in 1938 to explore the problem. These fruitful partnerships and the discoveries which bloom from them are not the product of chance. Discoveries are not only the achievement of the particular scientist, they are the product of their time. As the late Dr. Tatum put it in his Nobel address, a discovery depends on "knowledge and concepts provided by investigators, past and present all over the world; on the free interchange of ideas within the international scientific community, on the hybrid vigor resulting from cross-fertilization between disciplines; and last but not least on chance, geographical proximity, and opportunity."

In this team Dr. Beadle was the biologist, but a biologist who was not content with mere observation of biological phenomena. He yearned to intercede and try to understand biology at the level of molecular mechanisms. In particular, he was eager to probe the mode of action of a gene. While at the California Institute of Technology he joined forces in 1933 with a visiting French biologist, Dr. Boris Ephrussi, who was also impatient with the essentially descriptive approach of classical biologists to the basic problems of genetics. The two of them decided to explore a genetically controlled attribute of the fruit fly, its eye color. Normally the eye color of the fruit fly is brown, and this was known as the wild-type eye color, but by the mid-1930s some twenty-six other eye colors were known. These colors—vermilion, cinnabar, claret, etc.—are transmitted to offspring and must therefore be under the control of genes. Beadle and Ephrussi decided to transplant eyes from fly to fly in the larval stage before the colors had developed. With infinite patience they perfected their techniques until they could consistently create three-eyed monsters. Inspection of the developed trioptic creatures revealed an exciting discovery. The eye from a wild-type larva which was destined to be brown—had it been permitted to remain attached where it belonged—did not develop the brown pigment in its mutant foster larva. On the other hand, in the reverse case, an eye which normally would be vermilion developed brown pigment in its wild-type foster home.

Beadle and Ephrussi interpreted their finding with profound insight. They concluded that in the wild-type larva there must be some substance which can diffuse into the transplanted eye and convert it to the same brown color as the other two eyes. On the other hand, this pigment-forming substance must be absent from the vermilion-eyed host. That there must be some variants of the pigment-forming substance was shown when they found the following subtleties: a vermilion eye in a cinnabar host became brown, but a cinnabar eye in a vermilion host remained cinnabar. The investigators concluded that the inability of the flies with the exotic eye colors to develop the usual brown color must be due to a failure of their genes to produce the appropriate precursor for brown pigment formation. The subtle differences in the cinnabar and vermilion hosts were ascribed to sequential changes of the pigment precursor in the different mutants. These conclusions are still valid today. Dr. Beadle came to the conclusion that genes act by regulating chemical events. He therefore decided that he needed a chemist to extend his findings. He chose well.

Dr. E.L. Tatum had sound training at the University of Wisconsin in chemistry and biochemistry and in the role of vitamins in the nutrition of microorganisms. He also spent a year as a postdoctoral journeyman at the University of Utrecht in the laboratory of F. Kögl, who had just discovered biotin, a new member of the vitamin B group.

Dr. Tatum joined Dr. Beadle, who had by then moved to Stanford University. The first task to which they addressed themselves was a study of the nature of the eye-pigment precursor. Dr. Ephrussi had reported from Paris that if the amino acid tryptophan is incorporated into the diet of the fruit flies with mutant eye color there is a trace of brown pigmentation of the eye. Tatum took off from there and grew mutant fruit flies in the presence and absence of tryptophan under sterile conditions, and here is where Lady Luck waved her wand on Dr. Tatum for the first time. The tip of her wand was dipped in a culture of bacteria. The abode of one of Tatum's vermilion-eyed mutant cultures containing tryptophan became infected and the microorganisms altered the tryptophan in such a way as to render the amino acid an excellent source of brown pigment. Tatum isolated the substance produced by the bacteria and identified it as kynurenin, which is a decomposition product of tryptophan. This is as far as Beadle and Tatum went, but Adolf Butenandt, an excellent German organic chemist who was interested in the structure of natural products, carried the problem further and showed still another precursor of eye coloration—3-hydroxykynurenin. The sequence of enzyme motivated reactions and products turned out to be as follows:

Tryptophan
1
Kynurenin
2
3-Hydroxykynurenin
3
Brown eye pigment

The flies which can develop brown eyes normally can carry out all three of the above conversions. A vermilion-eyed mutant cannot carry out reaction 1. Therefore, if it gets kynurenin instead of tryptophan it can proceed to the complete pigment formation. A fly with the cinnabar eye coloration can perform reaction 1 but not 2.

Work with the fruit-fly eye color was frustratingly tedious for Beadle and Tatum.

A resilient flexibility in approach to a problem often distinguishes the great scientist from the merely competent. Beadle and Tatum decided to explore the mechanism of gene action in microorganisms. They were influenced in their decision by some penetrating deductions by an outstanding French biologist, Dr. André Lwoff, of the Pasteur Institute. Lwoff had studied the nutritional requirements of certain parasitic microogranisms, both protozoa and bacteria. He observed that many of these parasites often lost the ability to synthesize some of their essential nutrients, which they must siphon off from the hosts they infest. Lwoff clearly associated the loss of synthetic capacity with loss of the appropriate genes, for he spoke of the loss of some synthetic abilities as part of an evolutionary trend. And, of course, evolution operates through the selection of mutants.

The name of the organism Beadle and Tatum chose, Neurospora crassa, or bread mold, was recondite in 1940, but today it is a household word in biology. In making the choice the two investigators reached into their own scientific backgrounds and decided on this organism through an extraordinary set of coincidences. In the first place they wanted an organism in which the basic genetic work had already been done. Dr. Beadle recalled that while still a graduate student at Cornell he heard B.O. Dodge of the New York Botanical Garden give a seminar on genetic studies with Neurospora. Dr. Dodge kept on insisting that Neurospora is better than the fruit fly for genetic work. Tatum in turn brought a good deal of personal knowledge to the task. He had worked on the nutritional requirements of microorganisms, but more to the point he had shared a laboratory in Utrecht with Dr. Nils Fries, who had perfected a completely synthetic medium on which molds such as Neurospora could be grown. As we shall soon see this was an enormous asset.

The bread mold, Neurospora, makes very few demands on the world. It thrives on water, sugar, inorganic salt, and just a trace of one of the B vitamins, biotin. The reader may recall that Kögl, in whose laboratory Tatum spent a year, had already isolated biotin. It was available in pure form and thus there was no fear that unknown vitamins or other nutrients might sneak into the diet prepared for Neurospora.

The strategy of the experiment was simple. Out of the few starting materials in its diet Neurospora fashions everything it needs for growth of its cells. It makes its own amino acids, all twenty of them; it synthesizes all of its vitamins, except, of course, biotin; almost every cellular component of Neurospora is homemade.

The organisms were exposed to X or ultraviolet irradiation and were planted into culture media into which all of the known vitamins and amino acids had been incorporated, and the survivors of the irradiation were permitted to grow in this nutritional paradise. Spores of the organisms which thrived on the enriched medium were transferred to the original minimal medium on which their unirradiated ancestors—these are called the wild type—could grow with ease. To the delight of the investigators, some of the transplants did not thrive in the minimal medium: they had apparently lost, as a result of the irradiation, the ability to fashion some of their nutrients. Scores of new media were prepared in which the bare essentials were supplemented with a vitamin or an amino acid and the spores which were doing poorly in the minimal media were transferred to a partially supplemented environment. Several were found which thrived if only one amino acid or one vitamin was provided for them. And the requirement for the extra nutrient persisted after many generations. The experiment, so brilliantly conceived and faultlessly executed, was thus crowned with well-deserved success. Mutational changes were induced which were not expressed by gross alterations in the appearance of organs, as in the case of complex organisms, but by some invisible lesion within the irradiated cell whereby the capacity to synthesize a specific nutrient was lost. The irradiated Neurospora were crippled just as the irradiated fruit fly's offspring were, but in this case we knew exactly what was wrong. A mutation occurred that destroyed a specific capacity for a specific synthesis. But all chemical syntheses within the cell are operated by enzymes, therefore Neurospora revealed that a mutational change is a loss of some enzyme. The function of the gene now became clear: a gene must somehow give rise to a specific enzyme. Of course this is exactly what Garrod had stated some seventeen years earlier.

The second sweeping conclusion which was drawn from this

work is that genes, wondrous though their collective capability may be in creating a whole organism, be it mold, mice, or men, are individually highly circumscribed in their capacity. A particular gene can make only one specific enzyme.

Identical experiments were performed on bacteria, among which a great many nutritional requirements were known to occur naturally. The experiments were again successful beyond expectation. Bacterial mutants with nutritional requirements ranging from vitamins to nucleic acid precursor bases were soon produced by the scores.

It soon became apparent that there can be several different mutants that have the same nutritional requirement. For example, several mutants of a microorganism may be isolated, each of which requires the amino acid tryoptophan. But in one of them a compound called indole can replace the completed amino acid tryptophan. In another mutant an even simpler compound, anthranilic acid, will serve in lieu of the tryptophan. In a third mutant a still more primitive compound, shikimic acid, is adequate.

These various mutants revealed that a substance like tryptophan is not whipped together by a single knowledgeable enzyme. For potent though enzymes are, their versatility is invariably highly circumscribed. These mutants charted the plodding path of synthesis of tryptophan by a sequence of enzymes, each of which adds its bit to the total task.

There may be as many as a dozen sequential steps in the assembly line that makes a product such as an amino acid. Each of those steps is enzyme motivated and therefore a dozen genes must exist which carry the information for the shaping of those enzymes. If but one gene is lacking or is imperfect, the assembly of the amino acid cannot be completed. Should the missing enzyme be the one which completes the very last step of the synthesis, then the organism can survive only if the completed amino acid is supplied to it from an outside source. However, if the missing enzyme fabricates a part of the amino acid early on the assembly line, then any substance which comes after that step can substitute for the amino acid. Since all the other enzymes are present, they will work on the material which is proffered to them just as if it had been made by the missing enzyme, and thus complete the production of the amino acid.

Whether an organism can survive the loss of a gene depends on how pivotal is the gene's product, the enzyme, in the cell's total economy. It is interesting in this connection that, though many mammals have lost much of their amino acid and all of their vitamin-synthesizing capacity, all have retained through their long evolutionary ascent the ability to make all of the building units of their nucleic acids. In Nature only a few parasitic protozoa are known which lost some nucleic acid precursor-synthesizing ability. It would appear that these building units are too important to have their steady supply left to the uncertainties of the diet. Loss of these synthesizing capacities must have been too great a handicap, and such biochemical mutants must have been left as derelicts on the roadside in the surging path of evolution.

The inborn errors of metabolism which are clinically recognized are the result of gross aberrations within the genetic makeup of an individual. There are other gross differences in our genetic makeups which do not manifest themselves through any clinical symptoms. The reason is simply that the function of the enzymes involved is so peripheral that no disadvantage accrues from its loss, or gain. For example, there is a drug, phenylthio-carbamide (let us call it PTC for short), which, when placed on the tongue of a hundred humans will evoke a bitter taste in about sixty-six and absolutely no response in the others. Genetic analysis revealed that "tasting of PTC" is inherited as a Mendelian dominant trait. What that particular "tasting enzyme" is doing in the tongue of some of us we do not know, and its absence certainly cannot be classed as an inborn "error" of metabolism. There are finer fluctuations in enzyme levels among normal individuals. No two human beings are exactly alike. Differences in gross structures are obvious: identification by our fingerprints takes advantage of just one area of individuality. That we are individualistic even at the molecular level has become obvious from the work of Dr. Roger J. Williams of the University of Texas. He and his associates measured in human subjects a variety of metabolic parameters. These included components of the blood, exretion of products in the urine, and measurable reactions to certain drugs—including "tasting and nontasting" of some. From these measured entities a chart was prepared for each subject. This was in the shape of a star with lines of different length radiating from a central point. Each line represented a measured entity, the length of the line indicating the quantity. Thus a lot of information was concentrated in a small area and a sort of metabolic fingerprint was obtained for each subject. No two persons showed the same pattern, indicating that we differ not only in gross structures but even at the level of our molecular components. The only exception to this rule are identical twins, whose metabolic fingerprints are always essentially the same. Since identical twins originate from one cell and are thus genetically identical, and since it is only they who have similar metabolic patterns, we have evidence of genetic control at the molecular level.

Our genes determine not only whether we are tall or short, dark or light, but also how we metabolize our sugar or cholesterol, what is the level of hormones in our blood, how we react to nicotine—in short, they are the ultimate determinants of the many-faceted jewel that is life.

The man who was to determine the chemical nature of the gene, Oswald T. Avery, went into research as a refuge from the practice of medicine. A small, shy, sensitive man, he suffered doubly with each patient: from the sickness itself and from the frustration at having such a limited arsenal against the disease.

Once he was given the opportunity in 1913 to devote his time to research he made the most of it. He let nothing interfere with his work at the Rockefeller Institute in New York. He lived across the street from the institute, he never married, and he hardly ever traveled. He husbanded his time and energy to the point where he would not answer correspondence notifying him of the receipt of some award.

Avery worked with only two or three associates at a time, several of whom were to achieve preeminence in their field. He published but rarely, only when he felt the contribution was significant. Those were calm and happy days for a real scientist. There were no grants, therefore, neither grant applications nor progress reports to write. The rewards for the scientist in money, social prestige, and power were practically nil. The only rewards were the joy of work and the appreciation and esteem from one's peers.

Avery believed in the biochemical approach to the study of the attributes of purified components extracted from living organisms. His faith, insight, and experimental skill revealed to us the mechanism of storage of hereditary information. Avery followed up an observation of an English experimental pathologist, Griffith, who had stumbled on an extraordinary phenomenon. Dr. F. Griffith was a medical officer in the British Ministry of Health who was studying the difference in virulence of various types of pneumococci. He isolated single bacterial colonies from the sputum of patients with lobar pneumonia, grew pure cultures of the organisms, and studied their virulence by injecting such homogeneous bacteria into mice.

He noted that the pneumococci which, on special nutrient plates, form little mounds of cells that have a rough surface were harmless when injected into mice and those that form smooth, glistening mucoid mounds were virulent. To study whether a harmless culture may revert to virulence, he killed a batch of smooth, virulent organisms with steam and injected this preparation along with live, rough, attenuated organisms into mice. The concoction proved to be lethal. When Griffith performed a bacteriological postmortem he made an extraordinary observation. He found that the agents of death were live, smooth pneumococci which were teeming in the blood of the dead mice. In the animal, the rough, harmless cells were transformed into smooth, virulent ones. Griffith's explanation of his finding in 1928 may have been influenced by the preoccupation of scientists of that period with nutritional problems. He thought that the dead, smooth pneumococci broke up in the body of the mouse and furnished a "pabulum" which the live, rough ones utilized to build up a smooth structure.

But the molecular mechanism of this spectacular transformation in structure and virulence proved to be far more sophisticated than could be visualized by its discoverer. Griffith had actually observed a gene in action!

However poorly Griffith interpreted his findings, he should be given full credit for describing a completely new phenomenon of profound potential significance. It is not easy to dig in the garden of biology where scientists have already wielded their tools and to unearth something truly new. To observe and lift out something really new requires special gifts. It requires a humility before nature and a hauteur toward one's colleagues. Nature's clues must be revered, prevailing dogma rejected. The scientist who can pioneer must have insight, he must know how to design a clear-cut experiment; he must have sufficient self-confidence to be sure that what he observes is real, and sufficient mastery of the field to know that what he has found is new. And, finally, a dash of luck helps too.

It is not too surprising that the first step toward the chemical identification of the gene should have been taken not by a biochemist but by someone who was a biologist. Biological experiments are much easier than biochemical ones for the simple reason that it is the intact organism which completes the experiment after the initial challenge.

Griffith's finding came to the attention of Dr. Avery, whose first reaction was incredulity, for what Griffith was claiming was nothing less than transforming the organism from producing one type of complex sugar of the bacterial surface to another type. The "smooth" and "rough" textures are but the visual expressions of the kind of complex sugars the pneumococci elaborate.

Avery, nevertheless, asked one of his young associates to repeat Griffith's experiment and when the findings were confirmed, Avery went beyond Griffith's experiments; he could take extracts of the dead virulent bacteria and add them to the nonvirulent organisms

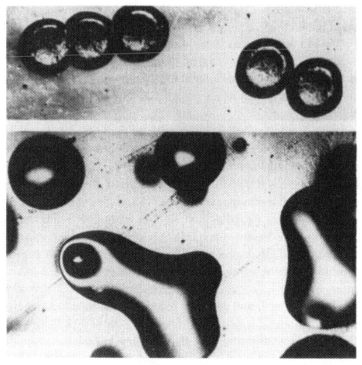

(PHOTO: PROFESSOR HARRIET EPHRUSSI TAYLOR)

Figure 10.1. In the upper panel are smooth, in the lower rough, pneumococci.

and convert them into virulent ones. Since these transformed bacteria, as they came to be known, would continue to reproduce and perpetuate their newly acquired attributes, they had undergone a hereditary alteration. Avery recognized that he had accomplished nothing less than an intrusion into the genes of bacteria, for it is in the genes that hereditary traits are enshrined.

A scientist on the frontier of knowledge is like an explorer who must make a series of decisions. Do I go on? Is there any promise or hope that further exploration may be fruitful? If he makes the wrong decision the explorer may risk his life; the scientist risks his reputation and, lately, the emoluments of a successful career.

Avery, intrepid explorer that he was, was not frightened by the awesome, implausible task he set for himself: the exploration of the nature of the gene itself.

Once a major decision is made, minor ones present themselves. If I go on, which way do I go? The directions are pointed by a vector made up of intuition and personal prejudice. Those who would have dared to suggest that the functional part of the chromosome is a single, identifiable entity would most likely have bet on its protein component as having the central role. The voice of logic dictated such a conclusion. Proteins are made up of chains with perhaps a hundred links of 20 different amino acid components. The permutations of variability are astronomical: enough to accommodate the multitude of known heritable traits.

Avery's own prejudice led him to favor some complex sugar as the genetic material, because of his knowledge that the shiny coat which was an attribute of the transformed pneumococci was composed of complex sugars.[2] In the early 1940s, it is safe to say, few if any scientists, among them Avery, would have put their money on another cardinal component of the chromosome, deoxyribonucleic acid, better known by its acronym, DNA, as the functional component of the gene. Though well-known and chemically well-characterized for decades, DNA was brushed aside as an unlikely candidate for a pivotal role in storing genetic information. This rejection was again dictated by the voice of logic: DNA's structure is too simple for the infinite capacity demanded of a reservoir of genetic information.

DNA, though vast in size, is very constrained in its component parts. A sugar (deoxyribose), phosphoric acid, and only four other components, the nitrogenous bases (adenine, thymine, cytosine,

2. As it turned out, the shiny coat contributed to virulence via a trivial mechanism. It protects the pneumococci against dissolution by the enzymes of the infected host.

and guanine), compose DNA, and in the early 1940s it was believed that even these possible sources of variability were equally distributed in different DNAs. Moreover, the intellectual spirit of the time was preoccupied with catalytic attributes, and in these, proteins are the performers without peer: some proteins can perform feats of chemical sleight of hand a million times a minute. No one could show any catalytic attribute of DNA. Indeed, the most educated guess about its function was that it was a reservoir of its components, such as phosphates, ready to release them as needed. At most it was thought by some that DNA is a supportive matrix on which the truly functional genetic proteins are stretched.

Avery patiently studied the extracts of pneumococci with their gene-changing potency. He isolated the known components—proteins, sugars—with the standard methods of chemistry. They were totally impotent in transferring virulence to the innocuous pneumococci. But the residue in the original extracts was still active. Chemical analysis of the residual component with the biological potency revealed it to be DNA.

Molecular evolution had a surprise in wait for us. DNA is inert and by itself it is as informative as the latest encyclopedia in the hands of Neanderthal man. DNA, to express itself, needs a system of translation involving literally dozens of proteins and three other kinds of nucleic acids. No one could have predicted this.

The importance of this flash of revelation was not lost on Avery, a man of boundless, but disciplined, imagination. However, its impact on the whole scientific world was that of a pebble in the ocean. The reason for this was twofold. Avery published his paper in 1944, when many scientists had much more compelling preoccupations than current advances in science: the world had to be secured free for science and other creative activities; indeed, for humanity itself. Moreover, communication in science was very limited. Basic science was still practiced as a hobby of a few very talented individuals. The amount of support from society or from governments, except for science with practical ends, was minimal. Avery's communication appeared in a medically oriented journal, and there were no vehicles as there are today which give notices of all articles in all journals. Today it is relatively easy to find out what was published last month in the *Journal of the Bulgarian Academy of Sciences*. As an example of how slowly information about undoubtedly the greatest single biological experiment of the century spread we might note that Avery, though he lived for 10 years after his discovery, which was to reorient all of our thinking, did not receive the Nobel prize.

But this is the price geniuses often pay for being too far ahead of their contemporaries; by the time their work is appreciated they may be gone. Yet they are not to be pitied, for with the ability to see what others miss often comes an inner vision, too, of their own worth. Avery *knew* the value of his discovery and he therefore passed the criteria of excellence for his most severe judge—himself. But more than this, he had the ultimate reward of a scientist. He saw in an exultant moment of transcendent brilliance what had lain deeply hidden through the eons of time since life began. He saw the blueprint of life.

11. The Spiral of Life

Creatures of circumstance is what Somerset Maugham calls one of his collections of short stories. His title refers to the denizens of a limited tropical world who are shaped by the triple circumstances of the tropics, boredom, and alcohol. But while his heroes—and anti-heroes—are limited, his description of mankind, and indeed of all things living, is universal: we are all creatures of circumstance. We have been shaped by the concatenation of chance forces which happen to prevail on this, our mother planet, in whose vast seas all things living had their origin.

Consider the very structure of water itself. That water is important to life is obvious. It covers three-fourths of the surface of the planet and it composes 70 percent of our body. But it is only the rarest chance which permits water to be the life-giving and life-shaping liquid that it is. If not for this chance attribute, water would be a gas at the temperatures now prevailing and our planet would be a gas-enshrouded, hot, arid, absolute desert devoid of any moisture and, consequently, of any life.

What is this attribute, then, to which we owe everything, the sweet climate of our planet and our very life? It is a very simple chemical propensity: the ability of hydrogen and oxygen atoms to affiliate with each other in excess measure.

It is a general rule that elements belonging to the same family behave quite similarly in chemical reactions. Thus oxygen and sulfur both form a dihydride. The dihydride of oxygen is, of course, water—the dihydride of sulfur is H_2S, hydrogen sulfide, the malodorous gas which issues from rotten eggs or from some sulfur springs. It is also a general rule that the heavier compound of such a homologous pair invariably has the higher boiling and lower freezing points. And here is where the unique attribute of water enters. H_2S has a molecular weight of 34, while water has one of 18, or the former has almost twice the mass of the latter. Yet the boiling point of water is 161° C above that of H_2S. Had water behaved as could be anticipated from the attributes of the other dihydrides of this family of elements, its boiling point would be not 100° Centigrade but −80°. And what a different world this would be as a consequence! In the first place the planet could not have cooled to its present temperature. It had been the repeated cycles over the millenia, of cascading rains and instant evaporation of the total contents of all the oceans, which have cooled our planet from a molten ball of lava to its present ambient temperature. Moreover, had water behaved as H_2S does, its vapors could not have condensed to form the seven seas until the average temperature had reached −80°. (This is colder than the inside of the most efficient of home deep freezers.)

It would have taken millions if not billions of years hence to reach such a low temperature, and once reached it would have been an inhospitable environment for the cumulative chemical reactions which eventually might have yielded that miracle of molecular evolution, a living cell.

To what do we owe the mild climate of our planet and the pleasant solidity of our bodies, 65 percent of which might have been a gas? We owe it all to the tendency of hydrogen and oxygen to form an unusual bond, the so-called hydrogen bond. Hydrogen sulfide (H_2S) exists as a discrete independent molecule and, therefore, it remains a gas until it is cooled to −61° C. Water, on the other hand, becomes associated with other water molecules by formation of extra bonds between the oxygen of one molecule and the hydrogen from a neighboring water molecule.

In this way, three molecules of water become condensed into one, forming a much larger and more sluggish molecule, which boils at 100° instead of −80°. Such a substance had the capacity to prepare our Earth for conditions which were propitious for the molecular evolution which led to life.

This, of course, is the most pervasive influence of hydrogen bonding, but as we shall soon see, hydrogen bonding plays a dominant role in the shaping of nucleic acid structures as well.

Avery's discovery, which revealed DNA to be the carrier of genetic information, slowly began to attract the attention of biochemists who flocked to the study of nucleic acid chemistry. They brought badly needed, newly developed tools with them.

All nucleic acids are very similar in chemical and physical attributes, rendering their separation tedious and uncertain. However, newly developed analytical tools soon permitted accurate analysis of the components of nucleic acids. The first to seize on paper chromatography, which we described earlier for the analysis of nucleic acids, were Dr. Erwin Chargaff of Columbia University and a Swiss postdoctoral visitor in his laboratory, Dr. Ernst Vischer. With this simple technique they opened up vast new frontiers of research in the origin and structure of nucleic acids.

DNA is built out of six major components. Four of these are organic molecules of varied structures: adenine, guanine, thymine, and cytosine. The first two and the second two are similar in general structure. The spatial volume occupied by each pair is very similar. However they are dissimilar in a very important atomic configuration; adenine and cytosine possess a similar appurtenance and so do guanine and thymine. All DNAs also contain the same 5-carbon sugar, deoxyribose, which gives DNA its generic name, and all DNAs contain phosphoric acid bound in organic combination.

Chargaff and Vischer discovered a regularity in the components of all DNAs. Adenine always equals thymine, and guanine equals cytosine in quantity. The meaning of this quantitative distribution eluded its discoverers but was soon comprehended by two other investigators.

Another tool brought to the study of nucleic acid structure is column chromatography. Dr. Waldo Cohn is a chemist who worked on the Manhattan Project, which developed the atom bomb. He studied separation of elements by what has come to be known as ion-exchange chromatography. This technique makes use of a highly charged granular material, usually some negatively or positively charged synthetic plastic, to pack a column. The mixture to be separated is absorbed on this column and is then eluted differentially by increasingly acidic—or basic—fluids. Hence the name "ion-exchange" chromatography; the ions which are absorbed onto the column are exchanged for the more strongly charged ions of the solution.

When the Manhattan Project achieved its goal, Dr. Cohn and his method suffered technological unemployment. But not for long. Being a very versatile gentleman—he is almost as good a conductor of symphony orchestras as he is a chemist—Dr. Cohn decided to apply himself to the study of nucleic acid chemistry. That he was permitted to do this is a monument to the wisdom of his superiors at the Oak Ridge Laboratories, for Dr. Cohn's switch of interest brought rich bounties in our understanding of nucleic acid structure. The most fundamental contribution Dr. Cohn made was the elucidation of how bases are strung together to build the giant chains which hold in their sequence the precious information for shaping a living cell.

Dr. Cohn showed that the four bases and the sugar components are attached to each other via phosphates to form chains of almost infinite lengths, forming molecules of huge sizes which were unknown in our previous experience with molecules of other substances.

How is genetic information encoded in DNA? How is DNA replicated as a cell divides so that identical genetic blueprints can be dowered out to both daughter cells? The answer to these questions which are uppermost in the mind of a biologist must reside in the structure of DNA. But there are more subtle questions as well: how are the genes, and therefore the DNA, divided in the formation of the sex cells? And once sexual fusion has taken place, how does the gene from one parent overpower its counterpart from the other sex cell to achieve dominance in the expression of a trait?

The classical biologists rarely attempted to correlate biological function with the structure of the cell's molecular components. For example, T.H. Morgan said in 1934 in his Nobel Prize lecture: "There is no consensus among geneticists as to what genes are—whether they are real or purely fictitious—because at the level at which genetic experiments lie it does not make the slightest difference whether the gene is a hypothetical unit or whether the gene is a physical particle."

But the biochemist cannot accept disembodied hypothetical units. He must strive to isolate such a factor and decipher its chemical structure, for it is an article of faith with him that the secrets of the mechanism of life are locked within the physico-chemical structures of the molecules that are the edifice of life. With respect to DNA the problem placed before us is the deciphering of a chemical structure which can explain, or is at least compatible with, its known biological functions.

For this task still newer and more subtly probing tools had to be applied to DNA. A school of X-ray crystallography was founded in England. With this technology, we can deduce the molecular configuration of crystalline substances by the distortion of X-irradiation which is shot through them. A student of this school, William Astbury, had the courage to tackle the forbidding problem of the X-ray crystallography of the macromolecules of the cells, the proteins and nucleic acids, even though the latter were not crystalline. As early as 1938 Astbury came to the conclusion that DNA has a linear, repeating structure in which flat planes are inserted at right angles to the long axis of the molecule.

Four new investigators took up the profoundly challenging jigsaw puzzle of the structure of DNA in the early 1950s. They are

M.H.F. Wilkins, Rosalyn Franklin, F.H.C. Crick, and J.D. Watson. Wilkins and Crick are English physicists who worked on the atom bomb project and on the design of aerial torpedoes and who at the end of the war also found themselves the victims—or rather, from their point of view the beneficiaries—of technological unemployment. A great many physicists chafe under conditions where their work is part of a group effort. If given a chance they run to biology with the eagerness of schoolboys running to a playing field after the day's restriction in the classroom, for in biology it is still possible to work alone. Wilkins and Crick searched and found a new area to engage their talents: the structure of biological macromolecules. Dr. James D. Watson is an American biologist who became frustrated with the limitations of the tools and ideas of classical biology and went to Cambridge to learn X-ray crystallography. Watson and Crick found each other intellectually congenial and began to poke around the problem of the structure of DNA.

The ease with which scientists join into working partnerships after the briefest of personal acquaintanceships comes as a surprise to those not involved in science. But this is a natural consequence of the nature of the profession: ours is a lonely pursuit. We live alone in a world of fantasy. No one else but those with training and imagination close to ours can become a friend, or a foe, in this fantasy world. I prize a close friendship of fifty years' duration with a distinguished lawyer. Whenever we meet he will ask in the middle of the first martini: "How's your work?" My answer: "So-so" or "Okay" disposes of that topic for the rest of the evening. But I can meet for the first time a Bulgarian colleague and despite the difficulty of expressing complicated thoughts in primitive, basic English, within minutes we are immersed in the intimate world of our ideas.

Watson and Crick had the following bits of the jigsaw puzzle to work with. Wilkins and his talented associate Rosalyn Franklin had repeated with the latest refinements the early work of Astbury on the X-ray crystallography of DNA. From the patterns of the dispersion of the X-rays, they came to the following conclusions. The

DNA molecule is made up of regularly repeating units, and the molecule is helical. The regularity of the repeating structures presented a paradox. Chemical analyses indicated a variable content of the component bases of DNAs from different sources, yet from this variability a repeating regularity emerged in the X-ray pictures.

This information became available to Watson and Crick, who were made aware of the analytical findings of Chargaff which indicated the quantitative equivalence of guanine and cytosine and of adenine and thymine. Finally the English biochemist Gulland emphasized the presence of extensive hydrogen bonding and had even suggested that the chains of DNA might be linked together by hydrogen bonds to form multichain complexes.

Watson and Crick now brought a still newer tool to the task: conformational analysis. This intellectual pastime is an outgrowth of the information about the structure of atoms and molecules which the X-ray crystallographers had gathered. The actual size of the various atoms is quite accurately known and the distances between atoms in molecules have been carefully measured. From this information we can build accurate scale models of atoms with little connecting tie rods that reproduce the appropriate bond distances. Scientific research has been facetiously called organized play for adults; molecular kits are our Erector Sets. With such atomic models the plausibility of the existence of various molecular structures can be explored. For if atoms cannot reach each other or are too bulky for a given site in the scale model, their stable existence in a molecule is also highly unlikely.

Watson and Crick made scale models of the four bases of DNA and they found that if they brought together (as if bonded together by hydrogen bonding) adenine and thymine and guanine and cytosine, the overall dimensions of these two couplets were identical. This was cause for rejoicing, for according to Chargaff's data the units in each of these couplets are present in equivalent amounts and if they formed such pairs in DNA, such analytical equivalence would be obligatory.

The formation of such hydrogen-bonded pairs with identical

overall dimensions would resolve a paradox presented by the X-ray data: the emergence of nearly identical repetitive structures out of the disorder of the random sequence of individual bases; the repeated units are not single bases of different sizes but base pairs of identical dimensions.

From Rosalyn Franklin's meticulous X-ray crystallography and expert calculations, it also became apparent that DNA is a helical coil with twists of identical size repeated to vast lengths.

The Watson-Crick model of DNA (as it is called) is, then, a double helix with each chain twisted around the other. The sequence of bases in each chain complements that of the other.

This brilliant hypothesis is consistent with all the known physical attributes of DNA and in addition it makes a prediction about an important biological function of DNA: the mode of replication of DNA itself.

Prior to cell division the DNA content of a cell must be doubled so that in each daughter cell faultless copies of the total genetic material can be invested. A structure of DNA with two helical chains fused by hydrogen bonding provides an ideal model for such replication. When the two strands separate as the two halves of a zipper, each twin coil serves as a model on which a complementary coil can be fashioned, i.e., a thymine vis-à-vis an adenine, and guanine vis-à-vis a cytosine, then the exact replication of the DNA and, therefore, of a gene is ensured.

Such a method of replication of DNA is represented schematically in Figure 11.1. This concept of the replication of DNA was revolutionary, for it assumed that the DNA acts as a template and a directive guide for some enzyme, inducing it to manufacture more of the template. Such an enzyme was unknown in our total biochemical experience. It required a man of unusual intellectual power to conceive it, of enormous self-confidence to undertake it, and of complete self-discipline to demonstrate its existence. Dr. Arthur Kornberg, now of Stanford University, was the man who was equal to this task.

Kornberg assembled the four building units of DNA, but intui-

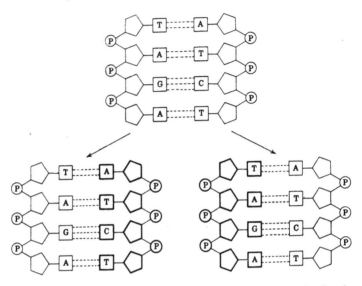

Figure 11.1. The birth of new DNA; the bold structures in the daughter chains represent new synthesis.

tively he chose triphosphates, which as with ATP have high energy stored in them. He then searched for an enzyme which in the presence of added DNA, to serve as a template, could fashion more DNA from the four building units. With patience and skill, Kornberg pursued the search and found an enzyme which could copy identical replicas of the DNA template proffered to it. The enzyme, DNA polymerase, was unique in our experience; it was the first enzyme found which needed a model which it then copied verbatim. The validity of the process has been confirmed by the production with this system of a DNA with biological activity, the DNA of a small bacteriophage called $\phi\chi$ has been copied so well, the product can be injected into bacteria, and then can reproduce the bacteriophage in the bacterium.

How is genetic information inscribed in DNA? Avery's discovery posed a conundrum. The dynamic functional parts of all living cells are the proteins. These are the components with catalytic properties which carry on the multitude of functions of the living cell. Individual proteins are composed of different permutations of the same twenty amino acids. Thus for their precise synthesis there must be information for the alignment of twenty different component amino acids. If these twenty amino acids are to be specified by some structure in DNA, there must be at least twenty different and variable entities in DNA itself. However, DNA, as we stated earlier, has only four variable components, the four bases: adenine, cytosine, guanine, and thymine. How can a four-component message system be expanded to encompass twenty different signals?

The problem is the same as that faced by the inventor of the Morse code. How can the telegraph, which has only two message components, a dot and a dash, encompass the twenty-six letters of the alphabet? The riddle was solved by two scientists almost simultaneously. Since there are only four variables in DNA and they must be stretched to the twenty different components of proteins, it can be done only one way. More than one component of DNA must spell out an amino acid, and thus the number of possible signal units in DNA can be extended. If two of the DNA components stand for one amino acid, then the number of possible permutations available within DNA is 4^2 or 16. This is, of course, inadequate, falling short by four of the required twenty. However, if three of these base components of DNA are signals for one amino acid, then the number of possible permutations of variability is increased to $4^3 = 64$, which is more than enough to accommodate all the known amino acids.

This is very primitive numerology. It was almost an insult that molecular evolution could not devise a different, perhaps more complex, system of signals. These could be the geometrical configuration of the various bases or combinations of them. However, thirty years after these primitive suggestions were made, we can state with confidence that this simple numerology is indeed the basis of the storage of genetic information. Three bases in DNA represent a single amino acid of a protein.

The double-stranded structure of DNA is a remarkable accomplishment of molecular evolution. For by this simple mechanism two things are achieved. The DNA itself is strengthened: if there is a stress on one of the strands cleaving some of the chemical bonds, the other one, the mirror image strand, which adheres, will support it until repair mechanisms arrive. But, more importantly, this mirror image structure assures the almost infallible duplication of the information entrusted to DNA as the cell divides. Each strand can serve as a mold on which the second strand can be shaped, thus assuring equal distribution of the information of the DNA and also assuring continuity. DNA in each of us today is a direct, continuing replica of the DNA of some primordial ancestor.

Since the same components comprise the structure of all DNAs, consequently the DNAs of all organisms are very similar. The only difference is the permutations of the four bases in which are encoded the sequence of amino acids for the large variety of proteins. The near identity of all DNAs and the universality of the genetic code are a potential menace to the preservation of the individuality of a species. Since all DNAs are so similar, how can the integration of a foreign DNA—say, from an invading parasite—into the DNA of the host be prevented? If there were no fail-safe mechanisms against such a calamity, a woman who harbors tail-propelled protozoa in her genito-urinary tract, as some do, might give birth to a mermaid.

Evolution, that great inventor, perfected an obvious but ingenious system for the protection of DNA: it is a species-specific branding of all DNAs and a surveillance system for the brands.

As with all newly revealed biological mechanisms, the system is obvious only in hindsight. In most cases, the revelation of a new insight into the wondrous complexity and simplicity of Nature comes as a surprise, even as a shock. But as Louis Pasteur said, "A scientist must have the capacity for astonishment." The species specific branding of DNA was discovered in my laboratory by my students, and we were astonished!

The reason for our astonishment was rooted in the path of development that the discipline of biochemistry had taken. The founder of modern biochemistry, Sir Frederick Gowland Hopkins of Cambridge University, pointed the way. When he decided to study the "chemistry of the living" at the turn of the nineteenth century he was cautioned by his presumably knowledgeable contemporaries: "the chemistry of the living, that is the chemistry of protoplasm; that is superchemistry; seek my young friend for other ambitions." So Sir Frederick decided to avoid superchemistry and contented himself with the study of small molecules of the living. He discovered vitamins and other small molecules present in all organisms. Biochemists who followed in his footsteps enlarged the inventory of the atlas of small molecules. From these studies emerged—as stated in chapter 3—the principle of the unity of biochemistry: all organisms utilize vitamins to make identical co-enzymes; many of the pivotal enzymes are the same in all things living.

Since we were steeped in the generalization of the unity of biochemistry, we were astonished to the point of incredulity when we discovered the safety device for the protection of the individuality of DNA. The device is a group of exquisitely knowledgeable enzymes which, like branding irons, stamp a species individuality on all DNAs. The brands, however, are the same. They are small atomic configurations known as methyl groups, and their frequency and position in the DNA have a species specificity.

Soon after the discovery of the addition of methyl groups to DNA, it was suggested by several colleagues that they serve as signals for the beginning or ending of genes—a sort of punctuation as it were. We were dubious about such a function. Our reservation was based on the compelling logic of biology; since bovine DNA contains a larger number of methyl groups, and thus more frequent punctuations than human DNA, our genes are either longer—and there is no evidence for this, otherwise we could not use the insulin of a calf for the alleviation of diabetes—or the art of punctuation was neglected with our progress to *homo sapiens.*

The species specificity of the distribution of methyl groups pointed to a function other than punctuation. I wrote in 1964, "The introduction of methyl groups into DNA undoubtedly alters the conformation of DNA, conferring upon it species individuality. Such structural individuality might render difficult the integration of foreign DNA (from some infecting parasite) into the DNA of the host and thus species specific methylation would serve as a guardian of DNA."

Such a function was unequivocally documented after a decade of effort by the Swiss molecular biologist Werner Arber. It turned out to be an extraordinarily ingenious system.

It was found that there are DNA-protecting enzymes which can scan the DNA for the branding unique to that species and if the distinctive brand it not at its expected site, the enzymes chop the interloping DNA. The simplest example of this is the following: at certain intervals along the DNA chain of *E. coli*, the structure in Figure 11.2 appears. If the two methyl groups on the opposite strands are missing, an enzyme cleaves the DNA in two—at the site indicated by the arrows—by breaking the bonds that hold the DNA chain together. Once the foreign DNA is thus fragmented internally, the smaller pieces can be whittled away by still other enzymes that scavenge for such small fragments.[1]

It is this surveillance system which prevents fertilization by the sperm of a species widely different from the recipient female. The Minotaur—which was the putative fruit of intercourse between a bull from the sea and Pasiphae, who was one of the ladies of the harem of King Minos of Crete—could only have been the product of the fervent imagination of some storyteller of olden days. However, in closely related species, the DNA modification is not sufficiently distinctive to prevent cross fertilization. The mule is the

---

1. Arber received the Nobel Prize for his work in 1978. It was well deserved, for he labored on, even though others who were presumably more qualified in biochemistry *proved* that he was wrong.

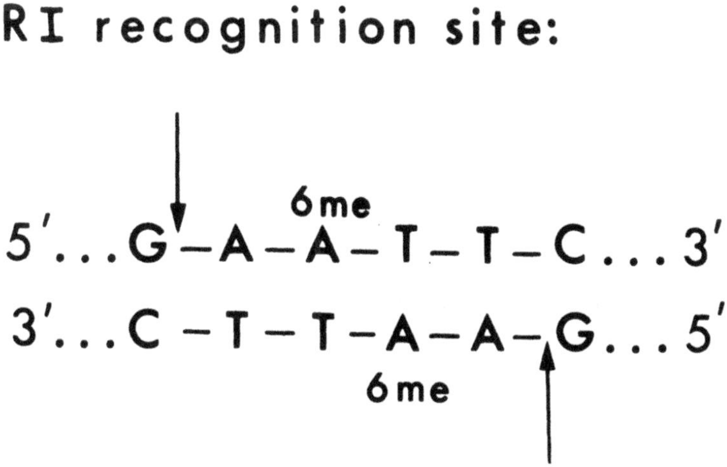

RI recognition site:

5′...G – A – A – T – T – C...3′
     6me

3′...C – T – T – A – A – G...5′
             6me

Figure 11.2. The Species-signaling Brands in DNA

product of cross fertilization by the ass and the horse. However, as is well-known, it cannot perpetuate itself; it is sterile. On my daily trip for lunch to my favorite restaurant, I pass by a butcher shop which sells "beefalo" meat, which is a cross between buffalo and cow.[2]

A whole host of these DNA-guarding "restriction" enzymes, which recognize the absence or presence of species specific passwords in the DNA structure, have been discovered. By choosing the appropriate enzyme, any DNA can be cleaved into pieces *in vitro.* The resulting fragments can be separated by sophisticated manipulations and compared to fragments produced by the same enzyme from a different DNA. A similar pattern of fragments implies the presence of passwords at similar spacings, and, consequently, the DNAs are from the same or very similar species. DNAs which are isolated from two viruses which are superficially identical in appearance and even in the antibodies they elicit can have different fragments when subjected to these searching en-

---

2. The steak looks and tastes like beef, but its texture is different, rendering it unappetizing.

zymes. Thus the DNAs of the herpes viruses, one of which causes the relatively benign infectious mononucleosis and another which causes the monstrous malignancies called Burkitt's lymphoma, do have somewhat different restriction fragments, which implies that portions of the virus DNA are similar and some portions dissimilar. Identifying the fragments that confer on one of these viruses its lethal potency awaits the skill and imagination of future researchers.

It was subsequently discovered that there are enzymes which can fuse *in vitro* restriction fragments of DNAs from the same species and indeed restriction fragments of DNAs from two different species. This is the method of producing "recombinant DNA."

The availability of recombinant DNA technology stimulated exaggerated fantasies of the eventual bounties that would grow out of the technology and even more exaggerated fantasies of its menace to mankind.

Fantasizing it easy; realization of fantasies is arduous and often impossible. The less experience one has in the latter, the more rampantly fanciful is the former activity. A molecular biologist at the high school level—and these days there are many of them—can pose tasks for recombinant DNA technology which would yield bounties beyond imagining: "If we would attach the DNA of nitrogen-fixing bacteria to the DNA of corn, the need for nitrogenous fertilizers would be forever eliminated."

Simpler problems have been posed and successfully completed. They were the synthesis by *E. coli* of two hormones, somatostatin, which is produced in the hypothalamus of our brain, and insulin. The original isolation of somatostatin required the brain tissue of half a million sheep. Since it contains but 15 amino acids, the DNA code for those amino acids could be synthesized. The technique in its simplest outline is as follows: there are small genetic structures present in *E. coli* called episomes, which reproduce independently of the main DNA of the coli. The episome DNA can be split by restriction enzymes and into the gap can be inserted the DNA which codes for the hormone. The adulterated episome can be reinserted into the *E. coli* which are made permeable by exposure to high concentrations of calcium ions. The coli proceed to transcribe and translate the foreign DNA to produce the desired hormone. The extraordinary achievement produced by a team headed by Dr. Herbert Boyer is a document to the validity of the concepts and technical achievements of molecular biology.

More recently a group of molecular biologists, including Dr. Boyer, has successfully completed a more ambitious undertaking: they cajoled the bacterium *E. coli* to produce insulin.

The remote possibility of dangers inherent in the manipulation of DNA in the laboratory was first pointed out by responsible scientists who were pioneers in the field. Had they been less conscientious in drawing it to the attention of their technically less well-informed but overly imaginative colleagues, the flap about the dangers of DNA splicing might never have been raised.

These fears stem from a marginal knowledge of molecular and descriptive biology coupled with an aptitude for unbridled fantasy. Of the myriads of microorganisms that share the surface of the earth with us, only a very tiny fraction molest us either directly or through some of their byproducts. Mutations among the benign, indeed beneficial, microorganisms occur with incalculable frequency, yet during the few centuries since accurate descriptions of symptomatology have been available, no evidence of new diseases has appeared.

Recombinant DNA processes may be occurring in our gut to a far greater extent than could be achieved by all the world's molecular biologists. Our intestines are teeming with billions of microorganisms. Of these, millions die every second of our lives; at the same time, the cells which line the alimentary canal, which are the fastest growing and consequently fastest dying of any tissues, are emptying their total content into the gut. The DNA-cleaving restriction enzymes as well as the DNA-sealing enzymes commingle in that vast laboratory of molecular biology, producing unimaginable numbers of permutations of spliced DNAs. Yet we are here. Although no DNA-guarding restriction enzymes have been unequivocally demonstrated in mammalian cells, because of technical difficulties, it is an article of faith with me that our cells are endowed with these to pounce upon any unusual DNA which may penetrate.

As a matter of fact, a man-directed DNA transfer, indeed genetic engineering, was performed some thirty-five years ago by Avery and his assistants. The conversion of innocuous pneumococci into virulent ones *was* genetic engineering. This was performed in the decrepit old laboratories of the Rockefeller Institute without any elaborate safety precautions, and no monstrous newly created pathogens have been observed crawling around E. 68th Street in New York City.

Actually some of our colleagues at medical schools who fear the imaginary consequences of man-made recombinant DNA may be exposed to greater dangers seeping out of the autopsy rooms. Our pathology residents have to perform autopsies on cadavers which teem with the infectious agents that caused the demise. During bone cutting with saws, they drip commercial hypochlorite—Chlorox—solutions on the electrical saws.

An appropriate incantation to Pandora, who let loose all of the diseases that debilitate us, might serve just as well as the hypochlorite drip.

Could we manipulate the total DNA of a human with sufficient finesse to reproduce multiple copies of an individual? The question evokes visions of horror in many minds. The thought is repellant to some on religious grounds, for we would be trespassing into the province of God, blaspheming the method of procreation ordained by Him.

Furthermore, the possibility of cloning a monster like Hitler, incubating a hundred clones from his morbid genetic endowment in a hundred eager freuleins is horrifying. But have no fear, dear reader, we are very far from this frightful achievement.

The term clone was originated by microbiologists; it describes a small mound of bacteria about 1–2 millimeters in diameter and half a millimeter high at its center, all of the bacteria being the descendants of a single bacterium which had been deposited on the semisolid agar some 24 hours earlier. Since all bacteria are the descendants of the same progenitor ancestor, they are presumed to be identical in their genetic endowment. However, the 10^7–10^8 cells clumped together are really not identical; mutations may have occurred which would destroy the genetic identity, and a variety of factors such as the diminished food supply in the center of the clone may cause transient lack of uniformity.

The term clone has, unfortunately, been transformed to describe a complex organism which is manipulated to life from a single cell without sexual fusion of the usual male and female sex cells.

Two scientists at the Fox Chase Cancer Center in Philadelphia, Drs. Robert W. Briggs and Thomas J. King, were the first to clone successfully frog cells to viability. The nucleus of an unfertilized amphibian egg is located just beneath the surface. Briggs and King made use of this anatomical happenstance for their manipulations. The nucleus can be pried out with a needle, or sucked out through an inserted microtubule; or the chromosomes in the nucleus can be destroyed by focusing a tiny beam of ultraviolet irradiation on them. Any of these manipulations leaves behind an essentially intact egg minus its nucleus. The next step is the insertion of a complete nucleus from any frog cell into the enucleated egg. Briggs and King hit upon a simple, ingenious technique for this transplantation into the foster egg. They used a tubule whose diameter was just too small to accommodate a frog cell but large enough for the passage of the nucleus. By sucking a cell from an intestine into such a tube they destroyed the integrity of the cell but not of its nucleus. The latter could now be inserted into an enucleated egg and anxiously observed—with fingers crossed. A small percentage of such eggs with living nuclear transplants can begin to divide just like a fertilized egg and can give rise to tadpoles.

Although these simple experiments, executed with expert skill, were impressive, their results could not be accepted as scientific proof of the presence of the total biological information for the individual in every nucleus. Devil's advocates (and some scientists, especially those whose ideas are destroyed by a new, ingenious experiment, take on this role with inordinate zest) might argue as follows: Are we certain that the information for the growth of the tadpole came from the transplanted nucleus? Could there not be some residual information left in the enucleated egg?

A biologist at Oxford University, Dr. J. B. Gurdon had the means to dispel any reasonable doubt about the validity of the conclusions. (Let us remember that the legal and scientific term "reasonable doubt" means a doubt entertained by a reasonable man. There are still a few scientists who reject these conclusions about the storage and utilization of biological information.) Gurdon had an indelible tag, a marker, on the transplanted nuclei which could be seen in the artificially created tadpoles. The nuclei of frog cells contain two distinct smaller bodies called the nucleoi. An observant graduate student at Oxford, Michail Fischberg, noted something strange in the cells of a line of South African frogs, *Xenopus laevis*. They had only one nucleolus! This is a genetic variation visible under an ordinary microscope which proved of immense value in answering a variety of questions.

Gurdon seized it to perform, among others, an impeccable experiment probing the amount of information in enucleated body cells. He repeated the procedure of Briggs and King; he destroyed by ultraviolet irradiation the nuclei of unfertilized eggs of normal *Xenopus laevis*. (These have, of course, two nucleoli.) Then he removed cells from the lining of the intestine of the mutant strain of the tadpole, which has but one nucleolus. Cells of the lining of the intestine are highly specialized, endowed only with the molecular apparatus needed for their appointed task: to assimilate the components of the food as it passes within their proximity. Since they are in eternal darkness, these cells need no visual sensory molecules; they can make no hemoglobin; they need synthesize neither pigments nor claws. They are stripped down for the most effective performance of their specialized tasks. But do they carry any more information within them than is absolutely essential? Have they, as they were stripped for their specialized functions, shed the information for making claws, vocal chords, and visual pigments?

Gurdon was soon to know. He destroyed the nuclei of about one hundred unfertilized eggs of normal *Xenopus laevis*. Into them he carefully inserted the nuclei from cells of the intestinal lining of the mutant frog with but one nucleolus. The frog eggs now contained nuclei endowed with the usual double dose of chromosomes. (The egg, of course, has only one half until the sperm delivers its quota during fertilization.)

Some molecular signal from the egg entered the nucleus and induced it to start yielding its hoard of precious information. A burst of molecular activity ensued which produced a doubling of vital components enabling the cell to divide, and the activity and cell division continued until a tadpole miraculously emerged. Only about 2 to 3 percent of such transplants were successful; but in these tadpoles the source of the information for their growth came unequivocally from the transplanted nuclei: every cell of the semimanmade tadpole contained but one nucleolus.

The reader can appreciate the technical obstacles which would need to be overcome in order to clone a mammalian DNA. To be sure, a writer whose cupidity is matched by his imagination claims to be privy to the creation of a human clone, but his document is so full of technical blunders, no knowledgeable scientist pays any heed to it.

It is just as well that cloning of a human is impossible at present because the product would be awfully lonely, as beautifully expressed by the writer Erica Jong in "Lullaby for a Clone (A Novelty Song for Our Age)":

> *How alone*
> *to be a clone*
> *to be the lovechild*
> *of the loins*

*of a love-affair with self
gestating on the shelf—
how alone to be a clone!*

*How alone
to be a clone
to be the duplicate of Dad
the total sum
of Mum,
how lonely & how sad—
how alone to be a clone!*

*How alone
to be a clone
how ghastly for
a blastula
to be both first and last of the
species of one—
how alone to be a clone!*

*How alone
to be a clone
to derive both
tooth & bone
from a family of one,
to be both father & yet son—
how alone to be a clone!*[3]

3. Copyright 1978 by Erica Jong.

Chance favors the prepared mind.

PASTEUR

12. The Weaving of a Protein

The largest and most intricate of manufacturing assembly lines created by man are puny compared to the assembly line which Nature has perfected for the creation of a protein molecule. The blueprint, of course, is the DNA, but it is as all blueprints, totally inert. To translate its contents with absolute fidelity to the desired edifice, a complete and perfectly functioning protein molecule, requires an assembly line whose complexity and efficiency fills those of us who know its intricate details with awe.

The physical chemist can only study multicomponent system interactions whose participants can be counted on the fingers of one hand. The molecular biologist must try to decode a mechanism whose components are numbered in the hundreds.

The fatuous riddle, "Which came first, the chicken or the egg?" has a counterpart at the molecular level: "Which came first, nucleic acids or proteins?" The analogy between the two is complete, for each riddle probes the origin of the components of a reciprocal, seminal relationship. The function of nucleic acids is to give rise to proteins; in turn, the function of some of those proteins is to give rise to more nucleic acids. We do not know the primeval order of the creation of these components of the machinery of life—nor shall we, unless a visit to Mars or some other planet may yield such a revelation—but we have learned a great deal about the almost miraculous interplay of nucleic acids and proteins in a contemporary living cell.

The first suggestion of the possible involvement of a nucleic acid in the shaping of protein molecules came around 1940 from the Belgian embryologist J. Brachet and the Swedish biologist Caspersson. It has been observed by histologists that rapidly growing tissue had great affinity for basic dyes and therefore such tissue must contain acidic components. Caspersson had designed a special microscope in which the optical components were made out of quartz rather than glass. The device was well worth its cost, for with such a microscope living cells can be observed not only in the visible range of light waves, but in the ultraviolet as well (glass is opaque to ultraviolet light). Nucleic acids are the only large components of the cell which absorb ultraviolet light intensively.

Therefore, under a microscope with quartz optical parts nucleic acid particles become pinpointed as opaque areas. Caspersson could thus identify the acidic components in rapidly growing cells as granules of nucleic acid.

Brachet made his correlation of nucleic acid levels and protein synthesis differently. For example, he observed that cells which make proteins for export are very rich in RNA. The silk gland of the silkworm, whose single-minded function is the production of a protein, silk, is the organ that is richest in RNA. On the other hand, the heart, an organ whose cells will make proteins only for repair, is one of the tissues with the lowest nucleic acid content.

The work of Caspersson and Brachet should have focused our attention on the involvement of nucleic acids—at least of RNA—in protein synthesis. But other ideas dominated the field at the time. In superficial hindsight it always appears that science moves from point to advanced point in a direct line to ever-increasing understanding of the world around us. This is a fallacy. It must be remembered that in science as in any other human endeavor, the participants are all too human. There are those who are ready to jump on a rolling bandwagon and bask in the security of working on a trendy hypothesis approved by the pillars of the profession. But there are those who, endowed either with real imagination or just with a perverse streak, will refrain from joining the crowd and will endeavor to hack out a path of their own.

The investigator who asked the most penetrating questions on protein synthesis and came up with answers with the strongest ring of truth is Dr. Paul C. Zamecnik of Harvard University. Dr. Zamecnik, a physician who turned to research, gathered around him an unusually talented group of young investigators at the Hungtington Memorial Hospital in Boston. (The caliber of an investigator's junior associates is one of the indices of his own quality. His mind, his personality, his whole being is exposed to the daily scrutiny of the most exacting judges, the younger colleagues who must use the association as a springboard for their careers.) Moreover, Zamecnik was particularly fortunate in having as a colleague on the staff Dr. Fritz Lipmann, an eminent biochemist who has had some of the most penetrating insights into molecular mechanisms. Dr. Lipmann had earlier deduced the source of energy in a living cell which drives the chemical processes that require energy: ATP. Lipmann proposed that ATP provides the energy for the weaving of a protein molecule from its precursor amino acids. Zamecnik and his coworkers soon confirmed this. They found that if the system which generates energy in a cell is poisoned with a drug, dinitrophenol, protein synthesis as measured by the uptake of a radioactive amino acid ceases completely. One of Zamecnik's associates, Dr. Philip Siekevitz, extended the frontier by showing that disintegrated cells, too, can incorporate amino acids provided various components of the cell are present in a relatively undamaged state.

Such a reassembly in a test tube of the components of disintegrated cells was made possible by the knowledge gathered by a group who were interested in anatomy at the subcellular level. The founder of this school was Albert Claude of the Rockefeller Institute, who, as so often happens in biological research, had something entirely different in mind when he initiated his search. Dr. Claude had become interested in seeing whether he could detect tumor-causing viruses. He chose to work with the Rous Sarcoma virus, which causes tumors in chickens. In his search for the cryptic virus, he developed methods of disintegrating cells and separating the various components by subjecting them to increasing gravitational forces in a centrifuge. The different fractions collected by this method were examined under an electron microscope and correlations were attempted with the components of whole cells. This work requires patience and visual imagination in equal measure. The organelles of a cell present entirely different images when they are clumped together in a mass and when they are integrated into the structure of a cell. The incidence of artifacts is high, errors of interpretation are inevitable. For example, it took considerable skill and imagination on the part of two investigators, Drs. George Palade and Keith Porter, to identify isolated clumps of pellets as structural components of the cytoplasm. These particles, which are called ribosomes, are globular structures rich in both RNA and protein.

Zamecnik soon found that the ribosomes were essential components of the protein-making apparatus; the radioactive amino acids in such cell-free systems were quickly concentrated into the proteins that were either integral parts of the ribosome or were adhering to it. The acidic particles which had been implicated in protein synthesis by Brachet and Caspersson were thus separated and identified at last as the ribosomes. However, the ribosomes alone were impotent in synthesizing protein. They had to be suspended in a soluble extract of the cells they came from, then, provided ATP was present as an energy source, the ribosomes could incorporate radioactive amino acids into proteins.

The reconstruction of components of disintegrated cells with sufficient finesse to achieve protein synthesis, however limited, was a signal achievement, although so far very little light had been shed on the intimate mechanism of the process. The "cell-free soluble extract," for example, may have hundreds of different, unrecognized components. As it turned out, such extracts became gold mines of biochemical information. The miner who wielded his pickax most effectively was another associate of Zamecnik's, Dr. Mahlon Hoagland. He had been trained by Lipmann in methods of detecting the flow of chemical energy from ATP. From studies of simple models it became evident that ATP forms a transient chemical alliance with the substance which is destined to be fitted into a new molecular pattern. It is as if for the synthesis of complex molecules the precursors stand on the shoulder of ATP and thus climb the wall to higher levels of energy. Hoagland was able to show that there are enzymes in the "soluble extract" that achieve a chemical union between ATP and amino acids.

By this time, in 1955 to be exact, the impact of the Watson-Crick model of DNA was beginning to be felt; the role of nucleic acids in determining the sequence of amino acids was becoming part of the intellectual atmosphere. Zamecnik and Hoagland began to explore the role of RNA in protein synthesis: is it just a passive template on which the protein molecule is stamped out or does the RNA of the ribosome serve a more active, a more involved function? To explore this question, they designed a complicated experiment from which emerged an unexpected finding of paramount importance. The favorites of the goddess of serendipity are those who dare to leave the neatly laid out paths of yesterday's research and make bold forays into the thick jungle on the far horizon. Some of these venturesome souls become hopelessly lost, but those who are lucky and alert—"Chance favors the prepared mind," said Pasteur—may be rewarded by sudden visions of beauty, as hidden truth reveals the answers to the questions they are seeking.

An unexpected message from their counter of radioactivity spelled out one of these truths for Hoagland and Zamecnik. The question they were asking was whether the reconstituted cell-free system which could incorporate amino acids into protein might also incorporate nucleic acid precursors into RNA. If so, RNA would be made simultaneously with protein. They added to one-half of their cell-free mixture the radioactive amino acid as before, and to the other half they offered a radioactive RNA precursor. A reagent called trichloroacetic acid—abbreviated TCA—can distinguish and separate RNA from protein. RNA dissolves in hot TCA, protein does not.

Therefore, if the "hot TCA soluble" fraction had been radioactive, it would mean that RNA was being made from the nucleic acid precursor. They found this to be the case, but to their great surprise they found something totally unexpected. The "hot TCA soluble" material was radioactive even in the mixture to which the amino acid had been added. Evidently the amino acid was being channeled somehow into the RNA. The surprised investigators confirmed and extended their findings rapidly. They found that the amino acid was bound to a fraction of RNA which had previously been known to exist, but whose function was totally obscure. But now a role for this RNA fraction had appeared at last. The amino acid was not yet bound into a protein but, rather, it was transiently associated with the RNA itself. The amino acid could be stripped away from the RNA by very gentle chemical manipulation. The RNA which accepted the amino acid, unlike the ribosomal RNA, does not sediment out, even if it is subjected to 100,000 times normal gravitational pull. The name soluble RNA, or "s-RNA," was given to this material which Hoagland and Zamecnik had recognized as being the conveyor of the amino acid during the assembly of a protein molecule. It should be pointed out that biochemistry had advanced by then to a sufficiently sophisticated stage in its development that predictions of mechanisms could be offered, provided there was enough insight and imagination. That a carrier of amino acids probably exists had been predicted by two investigators, Crick and the Belgian biochemist, H. Chantrenne.

An ingenious experiment revealed that soluble RNA is not a single substance but rather a complex one in which there are separate carriages for each of the twenty amino acids. This was revealed by observing that soluble RNA can be completely loaded with one amino acid until further addition of that amino acid does not increase the incorporation of radioactivity into the RNA. However, if a different amino acid is added, the previously loaded RNA accepts the additional burden. At this point still another amino acid could be loaded on.

Once the reality of soluble RNA was established, other investigators flocked to its study, and now, twenty-five years later, it is the nucleic acid about which we have the most detailed information.

Soluble RNA, now called transfer RNA—or tRNA—is relatively small, its molecular weight of about 25,000 is puny compared to that of DNA, which runs into the millions. The carriers of the different amino acids have been separated and were found to be indeed fastidious: each amino acid has at least one separate tRNA at its disposal which takes it to the site of the erection of the protein. The portaging nucleic acid molecules contain about seventy-five to eighty bases. The components at the terminal are the same. It is a sequence of two cytosines and one adenine. And we also know that the cradle into which the amino acid fits is a ribose molecule attached to the terminal adenine.

We can now write the first steps in protein synthesis in an abbreviated way as follows:

1. Amino acid + ATP Enzyme →ATP~Amino acid
2. ATP~Amino acid + tRNA-C-C-A→tRNA-C-C-A~Amino acid.

Or, if we prefer a visual image, every brick for the construction of an edifice of a protein molecule is brought in a separate wheelbarrow on a high platform and the brick is dumped onto the hod (the CCA end) of an individual hod carrier.

However, subsequent research revealed that this is but one of the roles of tRNA. It is the most versatile of biomacromolecules. For its many roles, evolution endowed tRNA with an extraordinarily complex structure. In addition to the four major bases, adenine, guanine, uracil, and cytosine, it may contain as many as twenty modified bases. The origin of these modified bases was obscure and baffling. Their mode of entry into specific positions in the tRNA chain could not be visualized by the base pairing hypothesis of Watson and Crick.

Fortunately, with the aid of extraordinarily good luck and dedicated work by two of my graduate students, Drs. Lewis R. Mandel and Erwin Fleissner, we solved the enigma of the presence in nucleic acids of modified bases which could not be specified by complementarity. About twenty-five years ago there was one microorganism which could guide us to the solution of this puzzle. Fortunately, it was part of a collection I acquired as a visiting scientist at the Pasteur Institute in Paris. This *E. coli* required the amino acid methionine as an essential nutrient. We addressed a certain question to the organism, but instead of answering, it revealed something else. When all other such microorganisms are deprived of an essential amino acid, they stop synthesizing all protein and all nucleic acids. However, this *E. coli* had lost all such self-control, and continued making RNA during the starvation of the amino acid. A couple of years later it was discovered by English scientists that some of the bases in tRNA contain methyl groups. It then occurred to me that since methionine is the source of methyl groups for most metabolites, there must be something wrong with the tRNA which accumulates during the methionine deprivation. Dr. Lewis R. Mandel showed by meticulous analyses that indeed the tRNA which accumulates during methionine starvation is lacking in its methyl group content. This tRNA became known as being methyl-deficient.

Another student, Dr. Erwin Fleissner, was then able to show that methyl-deficient tRNA could accept the full quota of methyl groups *in vitro* from certain enzymes. This finding then pointed the way to the origin of the modification of all nucleic acids.

The primary structure of tRNA was apparently insufficient for the many functions thrust upon it by evolution. And, therefore, a host of enzymes acted on the primary sequence of tRNA to modify it for its maximum efficiency. Some of these modifications are simple, such as the methyl group; others are extraordinarily complex. But there is no doubt but that all of these modifications are achieved by exquisitely knowledgeable enzymes, for they insert the appropriate modification only in specific positions. Moreover, the enzymes are species specific. In other words, the transfer RNAs in different organisms are modified differently. It was this species specificity which enabled us and others who entered the field to show the mechanism of methylation of DNA, for this, too, is highly species specific.

The species specificity of the methylating enzymes lead us to study these methylating enzymes in tumor tissue. We found them to be abnormally hyperactive in every tumor we examined, and this has been repeated throughout the world: there has been no exception. Every malignant tumor contains modifying enzymes of tRNA which are abnormal compared to the enzymes in the host tissue. In turn, it could be predicted that the transfer RNAs themselves in tumor tissue may be different and, indeed, this too turned out to be the case. There is no exception: in every malignant tumor there are a few transfer RNAs which are absent from the host organism. Only one of these transfer RNAs which is specific to tumor tissue has been analyzed for its methyl group content and, indeed, it was found in our laboratory that that particular transfer RNA contains an extra methyl group. Therefore, tumor tissue contains biochemical components which are *qualitatively* different from those in normal tissue. This is the first such qualitative difference in a tumor as compared to normal tissue. There is not more or less, of a transfer RNA, but a *different* transfer RNA.

We know from work with microorganisms that very slight modification of a transfer RNA can confer properties which can have a very large impact on the economy of the cell. What they do in tumor cells remains to be answered by future research.

Let us now return to the mechanisms of the assembly of a protein molecule. The blueprint for the building of the protein molecule, of course, must be the DNA. But the DNA is in the nucleus of the cell and protein synthesis occurs on the ribosome, which is a component of the cytoplasm. How is the information for the ordered array of amino acids transferred from DNA to ribosome? The answer came to two investigators who were asking a different question. Drs. Elliot Volkin and L. Astrachan were studying in 1956 the effects of bacteriophage infection on the cellular economy of the infected bacteria. These workers at the Oak Ridge National Laboratories knew from earlier reports of others that in bacteria infected by a bacteriophage, the synthesis of RNA essentially stops. But a certain amount of new RNA must be made because Volkin and Astrachan found that if radioactive phosphorus is offered to the bacteria right after the bacteriophage invades them, their RNA becomes radioactive. Since only the newly formed RNA contains the radioactive phosphorus, when the RNA is broken down to the component nucleotides—i.e., the base plus ribose phosphate—only the newly incorporated nucleotides will be radioactive. Therefore, the chemical determination of the amounts of the newly incorporated bases is easy; the amount of each base newly deposited is proportional to the radioactivity in the phosphate associated with it.

An arresting surprise emerged from these patient analyses. The RNA formed after the bacteriophage infection did not conform to the pattern of the four bases normally in the RNA of the bacteria. Instead, the bases mimicked the ratio of the bases in the DNA of the invading bacteriophage. Volkin and Astrachan concluded that a "DNA-like RNA" was being produced as a result of the infection. In other words, the bacterial virus, after invading the host, disrupts the normal processes in the cell and somehow orders an RNA to be produced that is the mirror image of its own DNA.

How is such a "DNA-like RNA" made in the cell? In 1959 Dr. Sam Weiss, a young man working at the Argonne National Laboratories in Chicago, designed an experiment fashioned after Kornberg's successful synthesis of DNA in a test tube. Dr. Weiss took DNA and an extract from bacteria and added to it the four precursor nucleotides of RNA. (Kornberg, of course, had used the four building units of DNA.) Dr. Weiss was able to show that the building units were assembled into RNA. The assembly of the RNA was totally dependent on the presence of DNA, for if the latter was excluded or destroyed, no RNA was produced. More than that, the newly fabricated RNA mimicked the base sequence of the DNA, which was used as the template. Now, a "DNA-like RNA" was produced not only in bacterial cells invaded by bacteriophage, but in a carefully reconstituted system in a test tube as well. Therefore, not only are there enzymes which achieve the duplication of DNA, but there are still others which can transcribe the sequence of bases of DNA into the complementary bases of RNA.

At this point—in 1961—two highly imaginative French scientists, Drs. Francois Jacob and Jacques Monod, both of the Pasteur Institute, synthesized these disparate bits of information into a unified theory of the molecular mechanisms of the genetic apparatus. Part of their contribution was the coining of vivid phrases for the processes and products involved, so that ideas could be easily verbalized and exchanged.

The first step in the mobilization of a gene into action is the "transcription" of the sequence of bases in the DNA into a complementary sequence of those bases in RNA. The RNA to which the genetic information is entrusted is a "messenger" RNA or mRNA. In the shaping of the mRNA the enzyme copies only one strand of the DNA. The alignment of the bases of the RNA is presumably by hydrogen bonding with the appropriate base pair of DNA. The one cardinal difference from the base-pair alignments in DNA is that, in forming RNA, uracil substitutes for thymine in pairing with adenine.

A scheme for the beginning of the transcription of one strand of DNA into a messenger RNA is presented in Figure 12.1 Once the transcription of the DNA into mRNA is completed, the latter peels off and is extruded—by we know not what force—into the cytoplasm. There it becomes attached to a ribosome and forms a complex with a tRNA which is waiting at the ribosome. The transfer RNA is attached to the messenger complex involving not only the RNAs but several protein factors as well.

For the formation of the complex, the ribosome is partially parted in two and into the gap the mRNA is inserted, where it moves along, much as a film moves along a ratchet. According to the original Watson-Crick theory, the RNAs with their appropriate amino acids find their specified positions on the mRNA and are

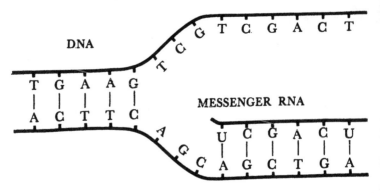

Figure 12.1. The Birth of Messenger RNA

locked into position by the complementarity of the three bases in mRNA with the matching three bases in the tRNA. Highly knowledgeable enzymes fuse the amino acids carried by adjacent tRNAs and from a series of such fusions emerges a protein molecule. To this process the term translation was assigned by Jacob and Monod. Many years of research effort was needed to uncover the minutia of the mechanism of translation, until today we are confident of its validity.

These were brilliant contributions which revealed the overall machinery of the extraction of information from DNA. However, all this still told us nothing about the actual sequence of bases which make up the code for each amino acid. At that time we were in the position of someone who heard signals, dots and dashes, coming in over a wireless in a more or less repetitive pattern, but did not know the Morse code and so was unable to decipher the message. The Morse code of the living cell was decoded by a young man, Dr. Marshall Nirenberg. As so often happens in science, this was not his original goal. Being a perfectionist, he cleaned up the system of translation so well that he could detect a chance, faint message, and thus it was he who broke the genetic code.

Dr. Nirenberg used essentially an artificially prepared messenger RNA which contained but one of the four bases. He had a messenger RNA with nothing but a long sequence of one of the bases, uracil. When he added this synthetic messenger RNA to his protein-synthesizing system, to his great surprise only one amino acid, phenylalanine, was used to weave an artificial protein molecule. What he obtained was a long sequence of phenylalanines attached to each other. Dr. Nirenberg immediately recognized the meaning of his finding. He had a repeated signal and thus the signal continued like a broken record repeating, in this case, the code for phenylalanine. Therefore, assuming a three-base code, three uracils specifies a phenylalanine.

Once the clue to the decoding of the first letter of the genetic alphabet was available, the rest was easy. It served as the Rosetta stone for deciphering all. Other synthetic messenger RNAs could be made by permutations of the four bases into the 64 triplet possibilities. Soon all the amino acids could be assigned an appropriate genetic code in the messenger RNA, and, in turn, in the DNA. To our surprise, for most amino acids there was more than one code. Again let us recall that there are 64 possibilities of permutations for the triplet code from the four bases and in turn there are only 20 amino acids. The significance of this multiplicity of the code did not become clear until later. Some of these codes signify that the amino acid is to be in a certain position of the protein chain. There were two triplet codes to which no amino acid could be assigned and, confirming the validity of the whole interpretation of this mechanism, we found that these orphan codes which have no amino acids assigned to them merely serve as punctuation. They do not represent any amino acid and, therefore, when they appear in the chain, their presence indicates that since no amino acid can be inserted to break the gap between the last amino acid and any subsequent amino acid, the message is terminated.

There are several other lines of evidence which confirm the validity of this triplet code. The most compelling one comes from the work of a brilliant biochemist, Dr. G. Khorana, who was working at the University of Wisconsin. He synthesized a synthetic messenger RNA which contained two bases in alternating sequence: ABA-BAB. If the triplet code has any validity, then this synthetic messenger RNA with the repeating two components could spell out two different code words, namely, ABA and BAB. Khorana chose the components A and B carefully so that ABA was the presumed code for one amino acid and BAB for another one. If the triplet scheme is valid, then in a protein-synthesizing system this synthetic messenger RNA should produce a protein containing only two amino acids in alternating sequence; and that is indeed what he found.

This was the crowning achievement in a whole series of brilliant experiments starting with Avery's discovery. It is a tribute to the capacity of the human mind and to the persistence of the biochemists and the molecular biologists who were not shackled by holistic mysticism and were, therefore, ready to explore the unexplorable. But we must reserve our awe for the extraordinary feats of molecular evolution which through eons of trial and error evolved this remarkable system. "Evolve" is possibly the wrong term. "Created" is more appropriate, and I use the term creation in its Biblical meaning. The genetic code had to be created anew. It has no counterpart in Nature. In the nonliving world there is not even the crudest counterpart or the most primitive forerunner of the storage of information. Catalysis by proteins has numerous facsimiles in organic and inorganic chemistry, but the specifying of a particular amino acid for protein synthesis by a code of three bases was an absolute invention: it was an awesome act of genius.

But, as far as we know, all of the information in DNA and consequently in RNA is linear information. All that evolution managed to accomplish in the storage of information is the synthesis of a linear sequence of amino acids. Why, then, is not every organism shaped like spaghetti? The reason lies in the wondrous ability of proteins to become molecular "erector sets" for the building of some very intricate three-dimensional structures.

The ability of proteins to provide shape is twofold. The first is catalytic. If an organism has a certain enzyme to make a mucoid polysugar, the shape and texture of its surface will be altered thereby. The second is inherent in the structure of the proteins: their component amino acids have a built-in propensity for folding and crosslinking, thus enabling them to shape interlocking three-dimensional constructions. There are examples of proteins folding into their preordained three-dimensional shape.

For example, the hormone insulin is composed of two strands of amino acids linked together in specific sites, as we have seen earlier, via a special atomic ability of one of its component amino acids, cysteine. It is easy to see how the sequence of amino acids to form each of these strands is inscribed linearly in a filament of DNA. But, once formed, how do these two strands find each other amidst the millions of other protein strands being stamped out at the same time on the assembly line of the ribosomes? Random collisions? The formation of so important a substance as insulin, whose absence dooms the organism to coma and certain death, could not be entrusted to chance. The following mechanism was evolved: the structure of insulin is linearly inscribed in DNA; but the inscription is about thirty amino acids longer than the actual structure of insulin calls for. The two functional parts of insulin are separated by an insert of innocuous amino acids. This insert can fold and bring the two precious parts in sufficient proximity to ensure the production of functional insulin by the formation of molecular bridges and by the elimination of the now superfluous insert.

The excess energy required to achieve this is enormous: to inscribe the information for the thirty extra amino acids, 180 extra bases must be woven into the DNA gene (three for each amino acid plus three to form the double strand of DNA). Ninety extra bases must be woven into the messenger RNA; thirty-one ligatures between the extra amino acids have to be formed. All of this—the synthesis of the bases, the synthesis of the amino acids, and their attachment into DNA, RNA, and protein—consumes energy. The squandering of this energy merely to ensure the meeting of strands A and B might appear wasteful; but any animal that tried to stint on this step is no longer with us. It disappeared for lack of insulin. Energy and molecular skill eliminate dependence on chance.

From Gurdon's experiment, then, we know that all the information of a species resides in the nucleus of every somatic cell. And we have just described the mechanism of the retrieval of that information by the synthesis of proteins from the stored information. The question we face today is: what is the mechanism of *selective* retrieval of that information so that only the proteins needed in a particular cell or at a particular stage in the development of an organism are synthesized? This mechanism, whatever it is, is the basis for the shaping of specific organisms via differentiation.

What is the molecular conductor which cues in and out segments of the DNA and thus orders the differentiation of each of the myriad of cells, endowing them with the wondrous variety of attributes essential for life's many tasks, and yet maintains those cells sufficiently akin to be amenable to integration into the harmonious whole that is an individual?

How does the molecular conductor falter and create a cacophony of DNA transcripts which are not amenable to integration into the harmonious whole and thus create the rampant monstrous growths of cancer? Our present concepts of these mechanisms of regulation are so crude, so lacking in biological reality and therefore so ephemeral it would be wasteful of the reader's time to relate them.

But once these now-hidden mechanisms are revealed and come within the reach of our tools, man may indeed become the master of his corporal fate. The practice of what can only be called molecular surgery might give us new limbs for old, may suppress deleterious genes from our seed, and enhance our creative capacity. Are these perfervid dreams? Perhaps they are to some; but not to those who ask, as Galileo did in 1615, "Who indeed will set bounds to human ingenuity?"

Time present and time past are present in time future.

T.S. ELIOT

13. The Next Chapter

The next chapter in the story is now being written in hundreds of laboratories all over the world: England, China, Russia, Hungary, and our country. The vast panorama of biological knowledge on which the curtain was lifted by American scientists with the aid of generous funding by the American people since World War II has become a goad for other societies as well. Scientists are trained, encouraged, and provisioned in all societies that can afford it. Since the demise and de-apotheosis of Stalin, the biological scientists are among the freest members in some of the People's Democracies. Provided they keep their political noses clean, they are permitted not only to travel to international meetings but also to spend as long as a year or more as journeymen in French, English, and American biological laboratories. They take back with them not only new skills, but new attitudes as well.

Meanwhile, paradoxically, in the fountain of scientific freedom, our country, the atmosphere is changing. There is an agonizing reappraisal—agonizing to the scientist—of the interaction of science and society. Such a reappraisal was bound to come; it is not entirely the result of financial strictures.

Convinced by the extraordinary success of our physicists in splitting the atom that science can pay immediate dividends, all echelons of our government endorsed enthusiastically the support of science and scientists.

Careers in science and the funding of science expanded steadily. We enlarged training programs until the demand exceeded the supply, and a few individuals with marginal motivation and ability were being guaranteed careers in science. I recall an almost painful shock when I first encountered two graduate students who frankly stated that they wanted to be scientists because of the security and comfort such a career offers. They were in for a rude surprise.

Funding of research was expanding at a rate which, if continued, could absorb the gross national product by the year 2000. Then the boom was lowered. Suddenly, severe cutbacks in funds for research and training resulted in utter chaos in the universities. The drain on the resources of some universities was so great they could not meet their commitments for salaries even of senior members of the staff, whose support from sources outside of the university suddenly vanished. The sad truth is that ladies' hairdressers have better job security in our society than do scientists.

Expectations of immediate beneficial results were fanned not by responsible scientists but by politicians. I was utterly depressed by the ceremonies of the signing of the Conquest of Cancer Act at the White House, which I unfortunately had the privilege of witnessing in 1971. It was clearly implied that since X billion dollars took us on a roundtrip to the moon, Y billion dollars would find a cure for cancer—where Y is much smaller than X. Now, nine years later, some of our politicians are demanding delivery on a promise we scientists never made, because we understood that an ultimate assault on cancer with our current knowledge of the biochemistry and molecular biology of a normal cell would be comparable to trying to get to the moon without knowledge of Newton's laws of gravitation.

The disenchantment of politicians with science and scientists is evidenced by the ridicule heaped by a certain senator on, to him, esoteric sounding research undertakings. Needless to say, he never heard of Pasteur's dictum: "Time alone can judge the worth of a discovery." Collectively, our politicians express their disenchantment by tightening the Federal purse strings.

The probability of funding a new biomedical research program from Federal funds shrank from about 70 to 30 percent or less.

Even well-established scientists await with bated breath decisions for the continued funding of their research.

If the continued productivity of many of our current scientists is uncertain, the careers of future scientists are in total jeopardy. The effectiveness of biological research in our country, whose general high quality is universally admitted, is due in large measure to the career opportunities in science for young people of modest means via the federal funding of their training. A number of studies have established that for some unknown reason, American scientists come from the lower economic echelons. Training of a biological scientist is long and costly; and a lucrative career such as that of a successful physician or lawyer does not await him. You cannot borrow on the prospects of such a career. Moreover, the thought of indenturing himself to the tune of $30,000 is frightening to a 22-year-old whose family has never had that kind of money at one time.

On the other hand, we are lauching crusades to conquer cancer, sickle cell anemia, heart disease, and everything else that ails us. But who will be the crusaders in the laboratory twenty years from now?

It is a pity, for there is much in the American spirit—independence, rejection of authority, tinkering ability—that has kept us in the forefront of biomedical science. If present trends continue, we may lose our preeminence. And there is much left to do. There is a minor school of biologists with an overly articulate apostle who claims that molecular biology is completed, it is a finished edifice; nothing new, just refinements of previous principles will emerge from its further development. This is nonsense. This was precisely what was said of physics at the turn of the century; only more refined measurements were left to be done. The brilliance of the atomic physicists of the subsequent decades illuminated the source of the defeatist resignation of classical physicists; their lack of imagination.

To fill accurately the vast gaps in our knowledge of the intimate mechanisms of control of differentiation, and of the initiating molecular forces which convert a potentially normal cell to malignancy, will yet take perhaps a century, or more, of dedicated, inspired effort by hundreds of molecular and cell biologists. Splendors invented by the genius of Evolution equal to those already unveiled are awaiting the questing minds and refined tools of biologists yet unborn.

What is needed is a stable, dependable level of support of both current biological research and the training of our best young people to roam the frontiers of biology decades from now. In the past few years such support appeared to be dictated by a roulette wheel rather than any guiding policy. Our federal funding of science is not unlike the finances of an old-fashioned family grocery store. You never knew from month to month what was available; at the end of a good year enough was accumulated for new shoes all around or even for a fur coat; in bad years there was nothing.

It is curious how readily a commitment of a score of billions of dollars was made to take our astronauts to the Moon, but how quixotic is commitment to the continuing exploration of our own biological world. The potential gain from the latter is infinitely greater than from the former. To be sure, as an adventure of the human spirit and mind nothing can excel our safe escape from our gravitional shackles, but the increment of scientific information—in spite of the spectacular claims made by the National Space Agency—which accrued from the many trips to the Moon is miniscule. We already knew infinitely more about the surface of the moon from exploration by earth-bound instruments than we know about regulatory mechanisms in a single cell in our little finger.

The decoding of secrets as yet hidden in our cells has inestimable potential value. Is there any doubt about the value of the discovery of insulin which has enriched and lengthened the lives of millions during the past half century? It is the firm conviction of many of us on the frontiers of biology that biological agents and mechanisms of equal worth and potency are awaiting discovery.

Among the latter we count the mechanism of brain function, which is admittedly beyond the reach of our current molecular and cell biology. Will we ever penetrate its mystery? Our leading Cassandra of biology rejects this challenge as being totally beyond our reach: evolution has not endowed our mind with sufficient power and finesse to penetrate this last arcanum. To be sure, progress is almost nil in this area. But that is no reason for a supine admission of impotence in the face of this ultimate barrier.

In the absence of knowledge one is prone to take refuge in mysticism. Until sentient man appeared, the splendors of Evolution were lost for want of an appreciative witness. Could a saber-toothed tiger value the triumph of the myriads of sequential mutations which produced it? Will the finest triumph of Evolution, the molecular mechanism of brain function, remain hidden and be forever unappreciated? One thinks not. Evolution, it seems, strove to create the uniquely powerful human mind to be appreciated at long last. Unlike all other creations of Evolution there was no gradualism in the emergence of the human brain. All other brains are puny and primitive compared to ours. If there be vanity in Nature she will not keep us at arm's length from solving this last mystery and admiring her ultimate invention.

Index

232 Index

Printed in the USA
CPSIA information can be obtained
at www.ICGtesting.com
JSHW021436221024
72172JS00002B/22

9 780231 043878